THE MONIST LIBRARY
OF PHILOSOPHY
SERIES EDITOR
Eugene Freeman
SAN JOSE STATE COLLEGE

Basic Issues

IN THE
PHILOSOPHY
OF TIME

Basic Issues In
THE PHILOSOPHY OF TIME

Edited by EUGENE FREEMAN
and WILFRID SELLARS

OPEN COURT • ESTABLISHED 1887 • LASALLE • ILLINOIS

BASIC ISSUES IN THE PHILOSOPHY OF TIME

EDITORS' FOREWORD

All of the essays in the present volume, except the essay by Adolf Grünbaum,[1] were first published in a special issue of the *Monist* (July, 1969), Volume 53, No. 3. This issue was devoted to a single general topic, "Basic Issues in the Philosophy of Time," and needed to be supplemented only by Grünbaum's essay to round it out into its present book form.

We are grateful to the authors of these essays, to the Open Court Publishing Company, publishers of the *Monist,* and to the Wesleyan University Press, for their cooperation in making these valuable essays available in the more permanent form in which they appear in the present volume.

Eugene Freeman and
Wilfrid Sellars

---

[1] Originally published, in a somewhat different form, as Chapter I of his *Modern Science and Zeno's Paradoxes* (Middletown: Wesleyan University Press, 1967); (2d ed., London: Allen & Unwin, Ltd., 1968); and reprinted in *Essays in Honor of Carl G. Hempel* (Dordrecht, Holland: D. Reidel Publishing Co., 1969).

CONTENTS

RECENT ADVANCES IN TENSE LOGIC*

1. *Lemmon's stratification.* By a "tense logic" I mean a system with the following features: (a) it contains sentential variables (say p, q, r, etc.) which stand for sentences (like 'Socrates is sitting down') which in some cases are true at some times and false at others; (b) it contains the usual truth-functions (say Cpq for 'If p then q', Kpq for 'p and q', Apq for 'p or q', Epq for 'p if and only if q', Np for 'Not p'), whose truth-conditions are given the obvious modifications, e.g. Np is true *when* and only when p is false, Kpq is true *when* and only when both its conjuncts are; and (c) it contains two additional functions (say Fp and Pp) which may be interpreted as 'It will be the case that p' and 'It has been the case that p', the former being true when and only when the plain p will be true later on and the latter when and only when the plain p has been true at some earlier time.

In my books *Past, Present and Future* (1967) and *Papers on Time and Tense* (1968) I gave some account of the main developments in this area that I then knew about; this article continues the story, mostly with notes on developments which either have occurred or have come to my notice since those books were written.

I shall, however, allow myself one major piece of recapitulation. Most of the developments I shall be discussing begin from the "stratification" of tense-logic which was achieved in 1965 by the late E. J. Lemmon. Lemmon divided tense-logical postulates into those which could be regarded as simply reflecting the obvious truth-conditions of tensed sentences, and those which could be regarded as reflecting specific postulates about the earlier-later relation, e.g. that it is transitive, that it has no first or last term, etc. If we write

---

*Editor's note: The editor is deeply saddened to report that he has learned from Mrs. Prior that Professor Prior died very suddenly in Norway, a fortnight before proofs of his article were sent to him. (The proofs were sent on October 13, 1969.) We are indebted to Professor Nicholas Rescher who has been kind enough to correct errors he found in the galleys.

Tap for 'It is the case at the instant *a* that *p*' and *Uab* for 'The instant *a* is earlier than the instant *b*', and use Π and Σ for the universal and existential quantifiers, the times when the different principal kinds of complex are the case are given by the following equivalences:

T1. *ETaNpNTap*   T2. *ETaKpqKTapTaq*

T3. *ETaFpΣbKUabTbp*   T4. *ETaPpΣbKUbaTbp*,

i.e. 'Not-*p*' is true at *a* if and only if *p* is not; '*p*-and-*q*' is true at *a* if and only if *p* is and *q* is; 'It-will-be-that-*p*' is true at *a* if and only if there is some moment later than *a* at which *p* is true; and 'It-has-been-that-*p*' is true at *a* if and only if there is some moment earlier than *a* at which *p* is true. T1 and T2 yield further equations for other truth-functions, and if we define *Gp* (for 'It will always be that *p*') as *NFNp* ('It will never be that not *p*) and *Hp* (for 'It has always been that *p*') as *NHNp*, T1-T4 yield the further equivalences

T5. *ETaGpΠbCUabTbp*   T6. *ETaHpΠbCUbaTbp*.

From these, the truth-conditions of complexes like *FHp*, *CFHpp*, *CpFHp*, etc. can be worked out, and certain tense-logical formulae can be proved from T1-T4 to be true at any arbitrary instant *a*. These constitute Lemmon's lowest "stratum" of tense-logical truths, his minimal system K_t. They turn out to be all those formulae which may be deduced by substitution, detachment and the rules

RG: $\vdash a \rightarrow \vdash Ga$ RH: $\vdash a \rightarrow \vdash Ha$

from propositional calculus plus the axioms

A1.1. *CGCpqCFpFq*   A1.2. *CHCpqCPpPq*

A2.1. *CpGPp*   A2.2. *CpHFp*

(Alternative axiomatisations are possible, e.g. *G* and *H* may be taken as primitive, *F* and *P* defined as *NGN* and *NHN*, and the A1's replaced by *CGCpqCGpGq* and its image. The A2's, also, could be replaced by *CPGpp* and *CFHpp*).

Further tense-logical formulae can be proved to be true at any arbitrary instant *a* by adding special conditions on the relation *U*. It is additions of this sort—the ones that belong to Lemmon's second layer—that form the subject of the advances that I wish first to record.

2. *New postulates for nonbeginning and nonending.* One condition that we may wish to put upon the earlier-later relation *U* is that it has no first or last term, i.e. that for any term a there is a term b which is earlier, $\Pi a\Sigma bUba$, and for any term a there is a term b which is later, $\Pi a\Sigma bUab$. (With these formulae as theorems we can of course drop the initial Πa, and assert them as $\vdash \Sigma bUba$ and $\vdash \Sigma bUab$). This addition to the minimal earlier-later calculus corresponds to the addition of the axioms $CNPpPNp$ and $CNFpFNp$ to K_t.

In earlier presentations of this result, at all events by me, it may not have been made sufficiently clear that these postulates mean no more than what has just been said, i.e. that the series of instants has no first and no last term. They do not express time's infinity in any quantitative or metric sense, i.e. they do not say that there is no maximum time *interval*, in either direction—they do not say that for any interval n, however large, every instant has an instant which is earlier than it by that interval, and every instant has an instant later than it by that interval. For the postulates of the last paragraph are quite consistent with the view that for each instant a there is some interval n such that no instant b is earlier than a by that much, though there are instants earlier than b by the whole continuous range of quantities *up to* (but not including) n, these instants having no earliest; and similarly on the other side. To distinguish this case from absolute quantitative infinity we would need a richer symbolic apparatus than the one so far described. On the earlier-later side we could do it if we had the form $Uabn$ for 'a is earlier than b by the interval n'. The weaker assertion of non-beginning would then come out as $\Pi a\Sigma n\Sigma bUban$, 'For every a there is an n and a b such that b is earlier than a by n', while the stronger assertion would come out as $\Pi a\Pi n\Sigma bUban$, 'For every a and for *every* n there is a b such that b is earlier than a by n'. If we define our original Uab as $\Sigma nUabn$, our earlier postulate $\Pi a\Sigma bUab$ will expand to $\Pi a\Sigma b\Sigma nUabn$, which is equivalent to the weaker assertion just made, while the stronger assertion has no exact equivalent in the dyadic U. On the tense-logical side, similar things are to be said about the relations between the forms Pnp and Fnp (for 'It was the case the interval n ago that p' and 'It will be the case the interval n hence that p') and our original Pp and Fp.

Endlessness without quantitative infinity is, of course, only possible if the series of instants is a dense one; in discrete time $\Sigma bUba$ and $\Sigma bUab$ *would* express quantitative infinity in the two directions. And whether time, if dense, has an intrinsic metric of a sort which makes the distinction between the two sorts of beginningless-ness and endlessness an intelligible one, I am not at all sure. At all events the distinction cannot be drawn within the nonmetric tense-logic and earlier-later logic with which we are now concerned.

On the tense-logical side, it has been obvious for some time that $CNPpPNp$ and $CNFpFNp$ are by no means the only formulae we could use to express time's lack of a first and of a last instant respectively. For example, they yield by transposition $CNPNpPp$ and $CNFNpFp$, i.e. $CHpPp$ and $CGpFp$, and these in turn yield them by transposition. These formulae, again, can be shown to be deduc-tively equivalent to $PCpp$ and $FCpp$, or to the rule that if a is any theorem then so are Pa and Fa. An intriguing variant of the latter, due to C. Howard (1966) is the rule that if Ga is a theorem so is a (this for no beginning), and the same for Ha (for no ending). Given Howard's first rule, we can pass from $\vdash a$ to $\vdash Pa$ via $\vdash GPa$ (this from a by $CpGPp$, and Pa from it by Howard's rule). And conversely, given that we may infer $\vdash Pa$ from $\vdash a$, we may derive $\vdash a$ from $\vdash Ga$ thus:—

1. Ga
2. $CCppGa$ $(1, CpCqp)$
3. $CPCppPGa$ $(2, \mathrm{RH}, \mathrm{A1.2})$
4. $CPCppa$ $(3, CPGpp)$
5. $PCpp$ $(Cpp, \text{rule for } P)$
6. a $(4, 5)$.

But there are certain other principles which have been known for quite a time to fail if time had a beginning or end. McTaggart, for example, in a work published posthumously in 1927 (*The Nature of Existence*, Vol. II, p. 20n.) having asserted that every event has "all three" of the determinations past, present and future, adds in a qualifying footnote that "If the time-series has a first term, that term will never be future, and if it has a last term, that term will never be past." (Here McTaggart's "will never be" seems to be doing duty for a tenseless 'at no time is') . Nelson Goodman, similar-ly, said in a work of 1951 (*The Structure of Appearance*, p. 366) that

A 'World War II was future'—if unaccompanied by any con-text determining what prior moment is being affirmed to precede World War II—says only what may be said about any event *that did not begin at the first moment of time.* Likewise, of any event *that does not run to the end of time,* we may truly say that it will be past. (Italics mine.)

So to say that *everything* that is the case "has been future," and to lay this down as something always true, is in effect to say that time had no past moment, and to say (as something true always) that everything that is the case "will be past," is in effect to say that time will have no end. This suggests $CpPFp$ and $CpFPp$ as axioms for nonbeginning and nonending respectively, and simple proofs of their deductive equivalence to $CHpPp$ and $CGpFp$ (given K_t) were found in 1967 by Storrs McCall. For the nonbeginning pair we have

1. $CpPFp$
2. $CHpPFHp\,(1\,p/Hp)$
3. $CPFHpPp\,(\text{CFHpp, RH, A1.2})$
4. $CHpPp\,(2,3)$

1. $CHpPp$
2. $CHFpPFp\,(1\,p/Fp)$
3. $CpPFp\,(2,CpHFp).$

3. *Strong and Weak Futures.* There is one law in Lemmon's minimal system K_t which doesn't *sound* as noncommittal as all that, namely $CpHFp$, asserting that whatever is now the case has always been going to be the case. By contrast with its mirror image $CpGPp$ (whatever is now the case will always have been the case), this thesis has for many a flavour of determinism. There are various ways of removing this impression. We might say, for instance, that if determinism is false there is, at each moment, no *actual* future but only a number of alternative *possible* futures, so that Fp can only mean 'It will be the case in some possible future that *p*'. $CpHFp$ would then mean merely that if anything has in fact come to pass then it must always have been "on the cards" that it would do so, and there is nothing deterministic about this. Genuine determinism would be the belief that there is only *one* possible future, and to express this you really do need to go beyond K_t and add a postulate for nonbranching of the future, e.g. $CPFpAApPpFp$ ('Whatever has been "on the cards" either is the case or has been the case or is "on the cards" still') .

If, however, in an indeterministic universe we read Fp thus, there must surely also be *another Fp* for which $CpHFp$ *would* have deterministic overtones—an Fp which means that p will be the case in *every* possible future. $NFNp$, i.e. Gp, is not exactly what we want here, for this goes too far; for since Fp means that p will be true *somewhere* in *some* possible future, *NFNp* would mean that it will not be true *anywhere* in *any* possible future that not *p*, i.e. that *p* will be true *throughout* every possible future. What we want is something in between, which says that *p* will be true *somewhere* in *every* possible future.

My own past attempts to bring out this distinction have been via a metric tense-logic. If *Fnp* means that it will be the case the interval *n* hence in *some* possible future that *p*, then *NFnNp* would mean that it will not be the case the interval *n* hence in any possible future that not *p*, i.e. that it will be the case the interval *n* hence in *every* possible future that *p*. In brief, if we read *Fnp* as 'It *could* be the interval *n* hence that *p*', we should read *NFnNp* as 'It *is bound to be* the case the interval *n* hence that *p*'. We may then define the weaker of the plain *Fp*'s as Σ*nFnp*, 'It could be the case at some future time that *p*', and the stronger of the plain *Fp*'s as Σ*nNFnNp*, 'It is bound to be the case at some future time that *p*'.

This last piece of English, however, is still ambiguous, and it is to be feared that the meaning captured by my formula is the wrong one, i.e. not the kind of "strong" or "definite" future in which people who operate with "definite" and "indefinite" future tenses are most interested. For Σ*nNFnNp* means that there is some specific interval *n* such that in every possible future *p* will occur after precisely that interval; and we have much less use for this than we have for the slightly weaker assertion that in every possible future *p* will be the case after some interval or other (*not* necessarily the same in each). We could only get this, one would think, if we had some way of quantifying over alternatives as well as over intervals.

In 1967, however, Storrs McCall cut this knot by digging back to the ordinary nonmetric earlier-later calculus, and putting forward a suggestion that amounted to the following: Use *H, G* and *F* as tense-logical primitives, with the truth-conditions of *H* and *G* given as usual by *ETaHp*Π*bCUbaTbp* and *ETaGp*Π*bCUabTbp*. Use *P* for *NHN* but do not use *F* for *NGN* but for a separate primitive whose truth-conditions are a little more complicated. If we write *Iab* for '*a* **is the same instant as** *b*' and *Bab* for *AAIabUabUba,* the latter will assert in effect that *a* and *b* are on the same branch. *TaFp* is then equated with

$$\Pi bCUab\Sigma cKKBbcUacTcp,$$

i.e. *Fp* is true at *a* if and only if every later **moment** *b* has some moment c on the same branch, and also later than a, at which p is true. That is, Fp is true at a if and only if on every branch issuing from a there is some instant at which p is true. This is exactly what is needed. The weaker 'It will be that p', meaning only that p is true at some later instant on *some* branch, is of course given by $NGNp$. It is now easy to assert $CpHNGNp$ but deny $CpHFp$.

McCall has suggested certain postulates for an *H-G-* (strong) -*F* system in which it is assumed that the earlier-later relation is transitive and that time is dense, infinite both ways, and non-branching in the past. The important and original part of McCall's list of postulates is, of course, the handful which involve the new *F*. These are

F1. $CGCpqCFpFq$ F2. $CGpFp$

F3. $CFpNGNp$ F4. $CPFpFPp$

F5. $CGNGNpGFp$ F6. $CFApFpFp$

F7. $CFpGAApFpPp$ F8. $CFAApFpPpAApFpPp$

F9. $CAApFpPpHAApNGNpPp$

F10. $CGAApNGNpPpAApFpPp$

Whether these are capable of further simplification, and whether they are complete, is not yet known. We may note that to prove Ta (F1) and Ta (F2) for any arbitrary a, nothing is required but the truth-conditions of G and F, i.e. F1 and F2 do not reflect any special conditions on U. (Whether the addition of F1 and F2 to K_t in G and H and will yield *all* H-G-F theorems which are thus independent of special conditions on U, I do not know). The presence of F2 in this "noncommittal" group is a little surprising, for with the ordinary F, $CGpFp$ only holds if time is endless. F2 and F3 do indeed yield syllogistically one of the usual postulates for nonending, $CGpNGNp$, but the "kick" in this comes not from F2 but F3 (F3, though not F2, is only true in nonending time). The reason is that when there is no future Fp, like Gp, is vacuously verified. For the condition of its truth at a is that *if* along *any* path there is a later instant than a, then at some point on that path it is the case that p; 'some' here is governed by an 'every', and what begins with 'every X' is vacuously verified when there are no X's.

If we add $CNGNpFp$ to McCall's system, this with F3 equates Fp and $NGNp$, and gives $CPFpAApFpPp$ (from F4 and F5) its usual interpretation as a postulate of nonbranching in the future.

4. *Density and circularity.* One condition which many would wish to impose on the earlier-later relation is that the series which it generates is a dense one. This condition is generally expressed by the postulate *CUabΣcKUacUcb*, 'If *a* is earlier than *b*, there is a *c* which is later than *a* but earlier than *b*'. This enables us to prove the truth at any arbitrary instant *a* of all tense-logical formulae derivable from K_t enriched by any one of the axioms *CHHpHp*, *CGGpGp*, *CPpPPp*, *CFpFFp* (weak *F*). If, however, time should happen to be circular, all of these postulates would be just as true if time were not dense but discrete. For in circular time every instant is both earlier and later than itself, i.e. we have *Uaa* as a law (and so *CUabKUaaUab*, and so *CUabΣcKUacUcb*), and whatever has always been or will always be true is true now (*CHpp* and *CGpp*, and so *CHHpHp* and *CGGpGp*), and whatever is true now both has been true and will be true (*CpPp* and *CpFp*, and so *CPpPPp* and *CFpFFp*).

How, then, *can* we distinguish discrete and dense time even if time should be circular? It has been observed by K. Fine and J.R. Lucas that since in circular time *U* relates any pair of terms at all, it is impossible for this job to be done with the symbolic machinery employed in the last paragraph; Lucas has also indicated, constructively, what further devices are needed (1966). In the first place, a *metric* earlier-later calculus or tense-logic will do; for in the former we can say that if ever *a* is earlier than *b* by the interval *n* then there is some *c* which is both later than *a* and earlier than *b* by some smaller interval than *n*, and in the latter we can say that if it will be the case the interval *n* hence that *p* then for some *m* less than *n* it will be the case the interval *m* hence that it will be the case the interval *n-m* thence that *p*; neither of these being automatically true in circular time.

Alternatively, without introducing metric conceptions, we may introduce into the earlier-later calculus a certain triadic relation among instants which is not definable in terms of any dyadic one. I shall not here use the particular triadic relation selected by Lucas, but another one which can be more directly echoed by an undefined dyadic tense-function. The relation I shall use is what might be called "directed betweeness," which must be distinguished both from undirected betweenness and from the directed betweenness which is definable in terms of earlier and later. If *a*, *b* and *c* are all points on a circle, each of them is in one way or another between the other two. If we speak of b being between a and c in a certain direction round the circle, there does seem to be a difference— Copenhagen is between London and Moscow if you're going eastwards but not if you're going westwards, and London is between Copenhagen and Moscow if you're going westwards but not if you're going eastwards. But if we define 'b is between a and c going east' as 'You can get from a to b and then from c, going east all the time', we don't capture this difference; for in this sense London *is* between Copenhagen and Moscow, even going east—we can go from Copenhagen right round the world eastwards (past Moscow) to London, and then from London round eastwards (past Copenhagen) to Moscow. If we define 'b is between a and c in the direction from earlier to later' as 'a is earlier than b and b than c', *this* "betweenness" becomes similarly vacuous in circular time; for in circular time we can say of *any* instants a, b and c that a is earlier than b and b than c. It is the *undefined* directed triadic temporal betweenness that we need, i.e. the analogue of the relation whose terms are London, Copenhagen and Moscow in that order, but not in the order Copenhagen, London, Moscow. Given this, we can assert density even in circular time by saying that whenever a is earlier or later than c there is a term between them in earliness or lateness.

Similarly in tense-logic we could introduce a form, say Tpq, for 'It will be that p and then q' in the sense in which this is *not* just short for $FKpFq$, "It will be that (p and it will be that q)"; and we could then assert density even in circular time by 'Whenever it will be that p then for some q it will be that q and then p', or by 'Whenever it will be that p, it will be that if-p-then-p and then p', $CFpTCppp$. We would, I think, want to have $CTpqFKpFq$ (though not its converse) as a law even in "noncommittal" tense-logic, and this with the preceding would give the more usual density principle thus:—

1. $CTpqFKpFq$
2. $CFpTCppp$
3. $CFpFKCppFp$ $(1, 2)$
4. $CFpFFp$ $(3, CFKpqFq)$.

5. *The logic of discrete future time.* As might be expected, axioms in P and F (or H and G) which are designed to express the view

that time is *discrete,* like ones designed to express the view that it is dense, fail of their purpose if time is circular. However, we now know what axioms in P and F (or H and G) do express time's discreteness, to the extent that that is possible with these primitives. A set of postulates for infinite, nonbranching, discrete time (with the earlier-later relation transitive) is given by R.A. Bull in the *Journal of Symbolic Logic* for March 1968 (not December 1967, as stated in *Papers on Time and Tense,* p. 161). There is no point in reproducing these postulates here, but it would be useful to axiomatise the purely future-tense part of this system, and I would conjecture that the following in G ($F=NGN$), subjoined to propositional calculus with substitution and detachment, would suffice:—

RG: If $\vdash a$ then $\vdash Ga$
G1. $CGCpqCGpGq$
G2. $CGpFp$ (for infinity)
G3. $CGpGGp$ (for transitivity)
G4. $CGCpqCGCpGqCGCFpqCFpGq$ (for nonbranching)
G5. $CGCGppCFGpGp$ (for discreteness).

Ordinary transpositions easily equate G5 with $CKFNpFGpFKNpGp$, which has been shown to yield (with the other postulates) the complete Diodorean modal logic (i.e. the modal logic with "Possibly p" defined as "It either is or will be that p") for discrete time. Moreover, when the mirror-images of these postulates are added, with the "mixing axioms" $CpPGp$ and $CpHFp$, we obtain Bull's complete past-future calculus for discrete time. The key proof is this:—

1. $CHGCGppHCFGpGp$ (5; RH; H1)
2. $CHGCGppCHFGpHGp$ (1; H1)
3. $CHGCGppCGpHGp$ (2, $CpHFp$)
4. $CHGCGppCGpHHGp$ (3, H3)
5. $CHGCGppCGpHPGp$ (4, H2)
6. $CHGCGppCGpHp$ (5, $CPGpp$) = Bull.

These facts make the sufficiency of RG and G1-5 for their purpose a likely conjecture, but as yet that is all it is.

Verbally, my G5 asserts that if it will always be the case that (if it will always be that p then it is the case that p), then if it will be that (it will always be that p), then, right now, it will always be that p. If time is discrete, this is true, since we can use the antecedent to come back step by step from the future permanence of p to its present permanence. But if time is not discrete there may be a future first moment of p's permanent truth but no last moment of its falsehood, so that the steps backward cannot begin. Except that if time is circular we have $CFGpGp$, and so G5, in any case.

6. *Yesterday and tomorrow.* Those features of the logic of time which underlie the *measurement* of temporal intervals have continued to be investigated by Hans Kamp. As was indicated in *Past, Present and Future*, pp. 106-111, Kamp operates with two undefined functions of which one means that at some past time q was true, and p has been true throughout the interval between then and now (leaving it open whether p is true now, and whether q has continued to be true since the time in question); while the other is the future-tense analogue of this. The past-tense one, on which we shall concentrate here, he now reads as "p since q," and symbolises as Spq. In 1965 he defined 'p the last time that q', and I later based on this an inductive definition of 'p the nth time ago that q'. Letting q mean 'It is midnight', 'p the last time that q' would mean 'p last midnight'. Kamp has now shown, in effect, how to define 'p at some time between last midnight and the one before', i.e. 'p at some time yesterday'. We first define 'p at some time today' as

'Not-q ever since (p-and-not-q, and not-q ever since q)', i.e. $SKKKpNqSqNqNq$. We may abridge this, in words, to 'p at some time since the last time that q' and in symbols to $P1pq$. We can then define 'p at some time between the last midnight and the one before' as

'Not-q ever since (q and (p at some time since the last time that q))',

i.e. $SKKqP1pqNq$. And in general, 'p at some time between the $(n—1)$ th midnight ago and the nth', or $Pnpq$, may be defined as

'Not-q ever since (q and (p at some time between the $(n—2)$ th midnight ago and the $(n—1)$ th))',

i.e. $SKKqP$ $(n—1)$ $pqNq$. Kamp's own 'p at some time yesterday' is not, indeed, quite the same as our $P2pq$, as he counts last midnight as part of yesterday; but his function is obviously obtainable as the disjunction of our $P2pq$ and 'p the last time that q'. Similarly his 'p at some time during the day before yesterday' is the disjunction of our $P3pq$ and 'p the last time but one that q'.

This is still quite a far cry from '*p* exactly $2\frac{1}{7}$ days ago', i.e. the *Pnp* of metric tense logic, but the further problems involved in reaching this goal can be expected to be increasingly those of the general theory of measurement rather than anything specifically tense-logical.

7. Predication and existence. All these developments concern tensed *propositional* calculi. In tensed *predicate* logic the main problem is, as it always has been, that of dealing satisfactorily with individuals which have existed or will exist, but do not exist now. This is one aspect of the more general problem of the relation between existence and quantification, a topic on which there is now a rapidly growing literature, with work by Cocchiarella, van Fraassen, Hintikka, Hughes, Lambert, Leblanc, Lejewski, Mates, Mayer, Routley, Scott, Thomason and others. I shall not go into details here, except to make an amendment to a system of my own, sketched on pp. 161-2 of *Past, Present and Future.* Here free individual variables are allowed to stand for nonactual or nonpresent as well as for actual or present individuals, but bound variables for actual or present individuals only, and there are two sorts of predicate variables—ϕ, Ψ, etc. for predicates quite generally, and *f*, *g*, etc. restricted to predicates which entail the existence of the individuals of which they are predicated. It is clear that with bound and free variables thus interpreted, the logic of quantification will not be quite standard, and we will not have either $C\phi y\Sigma x\phi x$ or $C\Pi x\phi x\phi y$ without qualification. In the passage cited, however, I give what purports to be a proof that we do have standard quantification where the predicates are of the restricted sort, i.e. we do have $Cfy\Sigma xfx$ and $C\Pi xfxfy$. Mr. A. Trew has pointed out to me (1967) that the second of these (though not the first) is counter-intuitive. The antecedent says in effect that whatever exists has a certain existence-implying predicate, which might very well be the case; but in the consequent *y* could be a *non*existent individual, who *wouldn't* have any existence-implying predicate such as *f*; this would give the alleged logical implication a true antecedent but a false consequent. In fact my alleged proof of $C\Pi xfxfy$ (though not that of $Cfy\Sigma xfx$) is invalid; it makes use of the rule ΠI, equivalent to the law $C\Pi x\phi x\phi y$, which does not hold in this sort of quantification theory. Benson Mates, in a formalisation of the thought of Leibniz, has a law restricted to *atomic* predicates which makes these existence-implying, and his laws for these work out the same as for my *f*'s.

I do not set much store by this system, as I am inclined not to regard sentences of the form *fy* as expressing propositions when nothing present or actual is named by *y*; but if systems allowing names of non-existents are to be developed, it is better that it be done properly. As I pointed out in *Time and Modality,* modal (tense-logical) systems which allow propositions of the form *fy* to be altogether absent from some possible worlds (at some instants) have to distinguish between 'necessarily (always) true' and 'not possibly (never) false'; and I sketched there a modal system Q in which this distinction was embodied. In the past two or three years this system has been very fruitfully worked upon by Dagfinn Follesdal in "A model theoretic approach to causal logic," *Dat Kgl. Norske Videns-kabers Selskabs Skrifter* 1966 Nr. 2, and by Krister Segerberg in "Some modal logics based on a three-valued logic," *Theoria,* 33 (1967), pp. 53-71, and *Results in Non-classical Logic* (Lund, 1968), Paper III. Segerberg contrives to fit Q very neatly into current schemes of modal semantics.

A. N. PRIOR

BALLIOL COLLEGE, OXFORD

APPENDIX: BIBLIOGRAPHICAL NOTE

At the end of my *Papers on Time and Tense* there is a numbered list of 47 articles, books, reviews, theses and mimeographed items on tense logic. To this list I make some additions below, with comments on some of the items. For reference, I mention first that item 20 on the previous list is my *Past, Present and Future.*

48. S. BLACKBURN, review of 61, *British Journal of Philosophy of Science,* 19 (Feb. 1969), 371-73.

49. R. A. BULL, review of 20, *Mathematical Reviews,* 36, No. 1 (July 1968), 10.

50. R. A. BULL, "Note on a Paper on Tense Logic," *Journal of Symbolic Logic*, 34 (1969), 215-18. Corrects 5.

51. J. BURIDAN, *Sophisms on Meaning and Truth* (Century 1966). A translation by T. K. Scott of Buridan's *Sophismata*, a fourteenth-century text well-known to contain much tense-logical material, especially in chapters 2, 4 and 5.

52. N. DARWOOD, "A Mathematical Notation for the Logical Delay," *Electronics Engineering*, 43 (1971).

53. R. GALE, *The Language of Time* (London: Routledge and Kegan Paul, 1968).

54. H. KAMP, "Tense Logic with Dates," UCLA multilith, 1967.

55. H. KAMP, "The Treatment of 'Now' as a 1-place Sentential Operator," UCLA multilith, 1967.

56. H. KAMP, "Discrete Operators in Dense Time," UCLA multilith, 1967.

57. G. KUNG, review of 20 and 61, *Philosophical Studies* (Maynooth, Ireland) 17 (1968), 237-45.

58. S. McCALL, review of 20, *Dialogue*, 6, No. 4 (March 1968), 618-621.

59. S. McCALL, "On What it Means to be Future" (abstract), *Journal of Symbolic Logic*, 33, No. 4 (Dec. 1968), 640.

60. A. N. PRIOR, last section of the article "Logic, Modal" in the *Encyclopedia of Philosophy*, ed. P. Edwards (London and New York: Collier Macmillan, 1957), Vol. 5, p. 12.

61. A. N. PRIOR, *Papers on Time and Tense* (Oxford: Oxford University Press, 1968).

62. A. N. PRIOR, "Now," *Nous*, 2, No. 2 (May 1968), 109-19. See also 67.

63. A. N. PRIOR, "The Logic of Tenses," *Akten des XIV Internationalen Kongresses für Philosophie*, Wien: 2-9 Sept. 1968, Vol. II, pp. 638-40.

64. A. N. PRIOR, "Egocentric Logic," *Nous*, 2, No. 3 (August 1968), 191-207.

65. A. N. PRIOR, "Fugitive Truth," *Analysis*, 29, No. 1 (October 1968), 5-8.

66. A. N. PRIOR, "Time and Change," *Ratio*, 10, No. 2 (Dec. 1968), 173-77. This corrects an error in 61.

67. A. N. PRIOR, " 'Now' Corrected and Condensed," *Nous*, 2, No. 4 (Nov. 1968), 411-12.

68. A. N. PRIOR, "Modal Logic and the Logic of Applicability," *Theoria*, 24, No. 3 (1968) 183-202.

69. A. N. PRIOR, "Tensed Propositions as Predicates," *American Philosophical Quarterly,* 6 (1969) , 290-97.

70. A. N. PRIOR, review of 53, *Mind,* 78 (1969) , 453-60.

71. A. N. PRIOR, "Worlds, Times and Selves," *L'Age de la Science,* forthcoming.

72. N. RESCHER, "Truth and Necessity in Temporal Perspective," in *The Philosophy of Time,* ed. R. Gale (New York: Doubleday, 1967), pp. 183-220.

73. N. RESCHER, *Topics in Philosophical Logic,* (Dordrecht, Holland: D. Reidel, 1968) , Chap. XII.

74. N. RESCHER and J. CARSON, "Topological Logic," *Journal of Symbolic Logic,* 33, No. 4 (Dec. 1968) , 537-48.

75. B. RUNDLE, review of 20 *Oxford Magazine,* Michaelmas 4, 1967.

76. B. RUNDLE, review of 61, *Oxford Magazine,* Trinity 4, 1968.

77. C. WILLIAMS, review of 20 and 61, *Ratio,* 11 (1969) , 145-58. This contains a comprehensive list of *errata* in both books.

78. G. H. VON WRIGHT, "Quelques remarques sur la logique du tempts et des systemes modales," *Scientia* 61 (Nov.-Dec. 1967) , 565-72.

79. G. H. VON WRIGHT, "The Logic of Action—a Sketch," in *The Logic of Decision and Action,* ed. Rescher (Pittsburgh: University of Pittsburgh Press, 1967), pp. 121-36.

80. G. H. VON WRIGHT, "An Essay in Deontic Logic and the General Theory of Action," *Acta Philosophica Fennica,* Fasc. 21 (1968) .

81. G. H. VON WRIGHT, "Always," *Theoria,* 24 (1968) , No. 3, 208-21.

PHILOSOPHICAL ASPECTS OF PHYSICAL TIME*

I. *The Universality of Quantum Time*

I would like to present a partial account of an investigation into scientifically and philosophically significant changes which quantum physics has made necessary in our views of time. In some cases, these changes resulted from discoveries of new aspects of time, as illustrated by the so-called "T.C.P. Theorem" due to Schwinger,[1] Pauli[2] and Lüders.[3] Their finding determines the transformation of the quantum state of any physical system resulting from a reversal of the direction of time, followed by a reorientiation of the dimensions of space and the replacement of each particle in the system with its antiparticle. A relativistic interpretation of the T.C.P. Theorem in Section III will show that it amounts to the universal interrelatedness of time, space and matter. In other cases, the changes in the scientific concept of time follow from the availability of new *methods* provided by quantum physics rather than from discoveries of unknown aspects of time. These methods may be applied to issues which bothered philosophers and physicists obsessed with the enigma of time long before Planck's quantum of action started our century. The quantum approach of von Neumann,[4] Pauli,[5] Fierz[6] and Lüdwig[7] to the second law of thermo-

*The research summarized in this paper was conducted under a grant of the National Science Foundation.

1 J. Schwinger, *Physical Review* 82 (1951).

2 W. F. Pauli (ed.), *Niels Bohr and the Development of Physics* (Long Island City, New York: Pergamon Press, 1957).

3 G. Lüders, *Zeitschrift für Physik* 133 (1952).
  J. Bernstein, G. Feinberg, and T. D. Lee, *Physical Review* (1965).
  Y. Nee'man, *Proceedings of the Israeli Academy of Sciences* (1968).

4 J. von Neumann, *Zeitschrift für Physik,* 57 (1929).

5 W. Pauli and M. Fierz, *Zeitschrift für Physik,* 106 (1957).

6 M. Fierz, *Helvetiae Physica Acta,* 28 (1955).

7 G. Ludwig, *Die mathematischen Grundlagen der Quantenmechanik* (Berlin: J. Springer Verlag, 1956).

dynamics and the associated irreversibility or asymmetry of time illustrate the use of the new methods for the solution of old problems of time. The relevance of quantum physical methods to the problem of reality of time will be shown in the last section of this paper.

In view of the complexity of the quantum physical issue of time, I shall limit this account to some aspects of the following, fundamental problems:

(1) The symmetry (reversibility, isotropy) of time and the associated invariance of laws of nature under time reversal.

(2) The relativity of time and the associated invariance of laws of nature under a Lorentz transformation of space-time coordinates.

(3) The physical reality of time and the associated empirical verifiability of laws of nature involving the concept of time.

To clarify the scope of these problems of quantal time, I shall first discuss the extent to which quantum physics restricts the meaningfulness of the idea of time and limits the solutions to the above problems which I shall suggest in this study. It is not the case that an investigation of time in quantum physics must be restricted to the physical microcosmos. At present, this marginal subuniverse of the physical world is supposed to consist of systems characterized by several orders of magnitude which approximate the physically meaningful minimum of any relevant magnitude. Thus, the size of a microobject is typically that of a molecule, an atom or an atomic nucleus, i.e., somewhere between 10^{-13} cm. and 10^{-8} cm. The "actions" involved in the microprocesses are N-tuples of Planck's quantum. The electric charges of microsystems are N-tuples of an electronic or positronic charge and the time intervals of the microprocesses may be of the order of 10^{-24} seconds, i.e., the time taken by a light-signal traveling over a path 10^{-13} cm. long.

The question of interest at this initial stage of our investigation is whether issues concerning quantum physical time are limited to the microcosmos. Such a limitation is obviously incompatible with the set of well-established quantum laws which deal either with objects of any ordinary terrestrial size, or with the physical systems and "supersystems" explored in astrophysics, cosmology and astronomy. The validity of quantum laws for physical systems of ordinary size is shown by the laws of superfluidity, superconductivity or of

the electromagnetic radiations corresponding to the hue and bright-
ness of colors visible with the naked eye. A most solid quantum
theory, viz. solid state physics, is in the same boat.[8]

The validity of quantum laws for oversized, cosmological systems
is apparent from H. Bethe's account of solar radiation in terms of
nuclear metamorphoses of hydrogen atoms in the sun into helium
atoms in the presence of carbon atoms. Another example is provided by the use of spectral analysis for the computation of the
cosmological "abundance" of chemical elements. In the nineteenth
century, the physicists who succeeded in determining the chemical
composition of several heavenly bodies by using this type of analysis
were as unaware of the quantum laws involved as Monsieur Jour-
dain was unable to realize that he had always spoken in prose.

Consequently, an investigation of quantum physical time must
not be limited to the microcosmos. Centering a study of time about
quantum physics means that time will be explored only within the
aggregate of physical systems which are governed by quantum laws.
Does the set of all such systems coincide with the set of all existing
systems, i.e., the universe? The apparent pervasiveness of quantum
laws suggests an identification of quantum physical time with
cosmic time. However, the conclusion which we shall eventually
reach in this study does not support a literally understood identification. We will conclude that quantum physical time is identifiable
with cosmic time, but has little in common with any intuitively
clear, prequantal idea of time. This conclusion will be shown to be
supported by the most revelatory quantum theory that is available
at present, viz. the theory of elementary particles. It is obvious that
the growing list of elementary particles[9] known to have several
properties with no classical analogue, including the strange property called "strangeness,"[10] records the most important experimental
discoveries of our time (which have always impressed the Nobel
Foundation and might have impressed Sir. E. Rutherford). Yet,
elementary particles are described in the most intuitive, classical

---

8 W. Weizel, *Lehrbuch der theoretischen Physik II* (Berlin: Springer Verlag,
1958).

9 J. Bernstein, *Elementary Particles and their Currents* (San Francisco: W. H.
Freeman, 1968).

10 M. Gell-Man, *Nuovo Cimento, Suppl.* (1956).

concepts of time.[11] The stability or instability of these particles, their life-span,[12] and their susceptibility to both destruction and creation are never doubted. The concept of collision[13] central to present research in quantum physics is usually construed as either synonymous with, or closely related to coincidence in time and space, i.e., the crucial concept of any relativistic theory of time and space.

The apparent overlapping of classical and quantum physical time is by no means a logical tautology. If this overlapping actually occurs, then one could hardly understand that the gap between the conceptual structures of classical and quantum physics is wide enough to make logically impossible the intertranslatability of these two groups of theories or the existence of a "language," i.e., a logically and physically interpreted formalism, that would be adequate for these two departments of physical science. At a later stage of this inquiry, we shall realize that this gap is not due to extraordinary changes of the orders of magnitude affecting all significant physical quantities and associated with the transitions from microsystems to standard sized, and, then, to cosmological systems and supersystems. Quantum physical time apparently has the unusual ability of surviving such transitions. The logical reasons for the pervasiveness of quantum physical time will be discussed in some detail in the closing section of this investigation.

Needless to say, it would be philosophically naive and logically premature to decide, at this juncture, whether or not we may consider the physicists' informal statements of quantum laws as evidence supporting the literal applicablity of intuitive concepts of time to the areas of the world which science seems presently to have conquered to some extent. There seems to be only one way to determine a philosophically and physically sound reinterpretation of concepts related to time. The point is that time has to be reinterpreted in the context of several physical theories, each physi-

11 P. Roman, *Theory of Elementary Particles* (New York: John Wiley & Sons, 1961).

12 T. O. Lee, R. Oehme and C. N. Yang, *Physical Review*, 106 (1957).

13 H. S. W. Massey, "Theory of Atomic Collisions," in *Encyclopedia of Physics* (New York: Reinhold, 1956), Vol. 36.

cal theory being a deductive system of universal laws.[14] Each law deals with an aspect of the entire universe. No physical law makes any particular, non-lawlike statements about a particular physical object, or a particular spatio-temporal region. Consequently, only a logical analysis of the set of physical laws which constitute quantum physics can lead to such a sound reinterpretation. More specifically, an analysis of the quantum physical laws related to the relativity, symmetry and reality of time can bring us closer to such an interpretation. This analysis will be attempted in the sections III through VI of this study.

In section II, I shall suggest an interpretation of the concept of law of nature, crucial in both physical science and its philosophy and involved in the three problems of time which we intend to explore. Parenthetically, let me first say for the benefit of those readers who, like me, may be concerned over the feasibility of our interpretational problems of quantum physics: they are not of the same type and do not raise the same difficulties to which R. P. Feynman[15] recently referred in an intriguing comment on the interpretational problems of quantum physics. He feels that such problems "are very difficult to state until they are completely worked out." The type of interpretational problems which he probably had in mind in suggesting a nonviciously circular and sound view of how to solve and to understand them can be illustrated by Born's interpretation of quantum mechanics which equates the expectation value of a physical quantity represented by a Hermitean operator A for a system the quantum state of which is describable by a square-integrable, complex valued function Ψ, to the scalar product of Ψ and the A-transform of Ψ. I do not know whether Einstein's suggestion to interpret quantum mechanics as a theory of Gibbsian ensembles of a specifiable type, now carried on by several theoreticians, e.g., Blokhintsev, would also fall under Feynman's class of interpretational problems. In the context of this study, no decision about the range of interpretational problems dealing with quantum physics, is either necessary or relevant. I have

---

14 H. Mehlberg, *The Reach of Science* (Toronto: University of Toronto Press, 1958) , pp. 202 ff.

15 R. P. Feynman and A. Q. Hibbs, *Quantum Mechanics and Path-Integrals* (New York: McGraw-Hill Book Co., 1965) .

to admit, however, that although the interpretational issues discussed in this paper can be shown to be outside of the set of interpretational problems exemplified by Born's statistical rules, the gist of Feynman's stand on the difficulties inherent in interpretational problems is not unrelated to the subject of this investigation.

One more remark is needed now to clarify the job ahead of us. The problems of the symmetry, relativity and physical reality of quantum-theoretical time can be formulated in two significantly different ways. Historically speaking, these two versions of the aforementioned problems of time have actually been proposed in several, important investigations. At present, one of these versions seems to have eliminated the other almost entirely, although there was no attempt made to justify the preference, nor could such a justification be carried out effectively. I have in mind what will be referred to in the sequel as either the *epistemological* or the *ontological* version of the symmetry, relativity and physical reality of time.

I am referring to the following bifurcation of physical and philosophical investigations into the nature of time: (a) The principle of *relativity of time* is now preferably worded as the requirement of Lorentz' covariance of all *laws* of nature. (b) The principle of the symmetry of time is normally replaced with the principle of the invariance of all *laws* of nature under time reversal. (c) The principle of the reality of time is now rarely referred to except in the writings of K. Gödel. What I am driving at here is that the ontological issue of the physical reality of time is associated with the epistemological question dealing with the empirical verifiability of the *laws* of nature involving the idea of time.

In the ensuing considerations, any statement ascribing to time the attributes of relativity, symmetry or physical reality will be designated as the "ontological" version of time relativity, symmetry or reality, in deference to those contemporary philosophers who classify under the heading "ontological" any issue dealing with "what there is."[16] On the other hand, the versions of the three quantum problems of time involving the concept of a law of nature will be designated as the "epistemological" versions of these prob-

---

[16] W. V. O. Quine, *From a Logical Point of View* (New York: Harper and Row, 1953).

lems in deference to all those philosophers of science outside France who include any problem concerning the functions of laws, facts, theories, etc., in human knowledge as "epistemological."

In French parlance, epistemology is synonymous with the philosophy of science while the problems which are classified under the heading of epistemology everywhere else would have to be reclassified as "gnoscological." Fortunately with regard to those specific quantum physical aspects of time I have listed the discrepancy between Anglo-Saxon and French linguistic preferences is irrelevant.

II. *The Concept of Natural Law Inherent in Scientific Time*

The concept of a law of nature is of decisive importance for the epistemological problems referring to quantum time; it is more frequently misunderstood in contemporary science and its philosophy than other central ideas.[17,18,19] I shall clarify this concept to an extent which seems necessary for this investigation. Let us begin with two classifications of laws of nature which, in conjunction, determine the subclass of laws of nature that can meaningfully be asserted or denied the type of invariance involved in all the problems considered in this paper. These two classifications will then be followed by an accurate, definitional explanation of the concept of *law of nature*. The need for such a definition will become obvious in the course of the investigation. I have been unable to find a satisfactory definition of this concept in publications pertaining to the *philosophy of science,* particularly in the well-known treatises of *Braithwaite,*[20] *Goodman*[21] and *Toulmin.*[22] It seems to me that the difficulties inherent in this issue are mainly caused by the interdis-

---

17 H. Mehlberg, cf. Ref. 14.

18 H. Mehlberg, in *Proceedings of the International Congress for Logic, Methodology and Philosophy of Science* (Stanford: Stanford University Press, 1960).

19 H. Margenau, *The Nature of Physical Reality* (New York: McGraw-Hill Book Co., 1954).

20 R. B. Braithwaite, *Scientific Explanation* (New York: Cambridge University Press, 1955).

21 N. Goodman, *Fact, Fiction and Forecast* (Indianapolis: Bobbs-Merrill, 1955).

22 S. E. Toulmin, *The Philosophy of Science* (New York: Harper and Row, 1953).

ciplinary role of the concept of a law of nature and by the lack of any *scientific theory* that could be used to cope with this definitional problem.

The first classification of laws of nature can be formulated concisely in terms of what will be shortly defined as the "logical rank of law of nature *l* pertaining to a physical theory T." The laws of T can usually be formulated in the so-called object language[23] L of T. However, in every advanced physical theory T there also are laws the formulation of which transcends the expressive possibilities of the object language L of T. Formulae for such laws are only available in an appropriate *metalanguage* L' of L. The idea of the object and metalanguage L, L' of a theory T are best explained for the cases when T includes a physically interpreted mathematical formalism. For instance, in *General Relativity*, the formalism consists, in the main, of a set of spatio-temporal variables supplemented by covariant, contravariant and mixed tensors of various rank, each of which is defined over the set of these variables. Some constant tensors are undefined within the deductive system but physically interpreted. *Einstein's* field-equations[24] involving the metrical and the energy-momentum tensors are expressible in the above formalism or the object language of this theory. Obviously, there are important laws in Einstein's theory which are not so expressible. In particular, his Principle of General Relativity and his Equivalence-Principle dealing with physically indistinguishable gravitational and metrical fields cannot be formulated in the mathematical formalism of the theory.

A metalanguage L' associated with the object language L must contain statements dealing with a set of significant logical properties of L. Thus, the classifications of the parts of speech in L ("logical types" in *Russell's*[25] terminology, closely related to *Lesniewski's*[26] "semantic categories") are describable in L'. Similarly, the *criteria of meaningfulness* of sentences of L, i.e., the condi-

---

23 R. Carnap, *The Logical Syntax of Language* (New York: Harcourt, Brace, and Co., 1937) .

24 A. Einstein, *Annalen der Physik,* 49 (1916) .

25 B. Russell, *Introduction to Mathematical Philosophy* (New York: Humanities Press, 1919) .

26 S. Leśniewski, *Fundamenta Mathematicae,* 14 (1921) , pp. 1-81.

tions to be fulfilled by those finite sequences of symbols of L which are meaningful statements in L must be expressible in L'. A language L'' which is interpretable as a metalanguage of L' can be said to be a second order metalanguage of L. A metalanguage of order N for L, where N is any natural number, is similarly definable.

From a logical point of view, we must guarantee the inequality L $\neq$ L'. Otherwise, the theory T would become antinomial or intrinsically contradictory. Such a logical sickness of T can only be cured by a crippling surgery which will enable a part of T to survive by eliminating another part of T, although the latter may be logically unobjectionable. The most effective and least expensive way of obtaining L $\neq$ L' consists in assigning the part of L' to what A. Tarski[27] calls the *semantic metalanguage of T* on the assumption that L is the formalism of a mathematical, deductive system. So far, the applicability of semantic metalanguages to object languages of physical theories has not been investigated. In a companion paper the reasons for both the validity and necessity of associating a semantic metalanguage with an axiomatizable or axiomatic physical theory T are specified.

The logical rank of a law of nature 1 pertaining to the theory T is $= 2$ if 1 can be expressed in the semantic metalanguage L' of T. The rank of a law in T is 3, if a formulation of this law is already available in L''. Laws of rank N can be similarly defined for any natural N. Suppose that T is quantum electrodynamics[28] whose fundamental laws result from a superposition of Dirac's equations of relativistic quantum mechanics on Maxwell's electromagnetic field equations formulated in terms of the vector and scalar potentials of the field. Furthermore, the Dirac spinors and the electromagnetic potentials are reinterpreted as operators over the Hilbert space associated with the compound quantum system made up of the electromagnetic field and a set of electrons and positrons obeying Dirac's laws. The logical rank assigned to the fundamental equations of quantum electrodynamics is then obviously $= 1$ whereas the statement of the relativistic covariance of these equations has rank 2, etc.

---

27 A. Tarski, "Der Wahrheitsbegriff in den formalisierten Sprachen," in *Studia Philosophica* 1 (1936), pp. 261-405.

28 A. I. Akhiezer and V. B. Berestetskii, *Quantum Electrodynamics* (New York: John Wiley & Sons, 1965), pp. 253 ff.

Similarly, von Neumann's quantum theory of measurement can be assigned the logical rank $= 1$ and construed as an integral part of his axiomatic quantum mechanics provided that the former be expressed in the proper, semantic metalanguage. By so defining the content of quantum mechanics we would come to a slightly pedantic but consistent reformulation of a definition recently suggested by E. Wigner.[29]

The classification of laws of nature in terms of logical rank must be supplemented by an essential subdivision: Some laws of rank 2 in T are indispensable in the sense that the omission of any such law would prevent the deducibility of a set of significant laws of nature from the remaining fragment of T. Such laws of rank 2 in T will be called "constitutive." They are exemplified by Einstein's two Relativity Principles, Wigner's[30] "superselection rules," etc. Laws of rank 2 in T which are not constitutive will be termed "extrinsic." W. R. Hamilton's[31] momentous analogy of wave optics and analytic dynamics is a case in point. Other examples are provided by von Neumann's law of rank 2 establishing the expressibility of quantum mechanics in the theory of Hilbert spaces and Wightman's[32] law of rank 2 concerning the expressibility of the same physical theory in the mathematical formalism due to L. Schwartz and usually designated as the theory of distributions.

The second classification of laws of nature is tripartite and can be put as follows: (a) Laws of nature which deal with a specifiable set of physical systems and can be expressed in terms of operator-valued or "c-number" valued functions defined over a set of space-time coordinates, are *intrasystemic*. Newton's laws of motion and Maxwell's field-equations of the electromagnetic field fall under this category. In both cases, the space-time coordinates are supposed to involve an inertial frame of reference. In general relativistic theories, the set of frames of reference relative to which the laws of nature are valid, is more comprehensive. (b) Laws of nature dealing

---

29 R. M. P. Houtappel, H. Van Dam and E. P. Wigner, *Review of Modern Physics,* 37 (1965).

30 E. P. Wigner, *Group Theory* (New York: Academic Press, 1959).

31 W. Weizel, Ref. 8, I (1955).

32 A. S. Wightman, "Quantum Field Theory in Terms of Vacuum Expectation Values," *Physical Review,* 101 (1956).

with functional interrelations among various coordinate-systems and with more general transformation-properties of the complex aggregate of physical quantities are *intersystemic*. The law asserting the validity of the Lorentz group of transformations is a case in point. Other significant examples of intersystemic laws deal, e.g., with the transformation of those Hermitean operators over a Hilbert space which are associated with a physical system and represent measurable, physical quantities under an (infinite dimensional) unitary and irreducible representation of the Poincaré group. My references to the Lorentz and Poincaré groups and their representations, instead of Larmor groups, is justifiable on grounds of scientific conformity. (c) Laws of nature are called *extrasystemic* if they involve no frame of reference at all. The laws stating the atomicity of electric charge or the four-dimensionality of the space-time continuum fall under this category.

In the context of our investigation the important point is that no invariance requirement can be meaningfully applied to a law of nature unless it is intrasystemic and of logical rank 1. Accordingly, invariance laws involving the Lorentz transformations or time reversal would be meaningless. Incidentally, the definition of a law of nature recently proposed by a well-known physicist[33] and requiring that the set of laws of nature should be identified with the set of all physical statements which are invariant under the transformations of the Lorentz, the gauge and the canonical group is obviously too restrictive.

I shall now proceed with constructing a definition of intrasystemic laws of nature of rank 1 on the basis of the two proposed classifications of laws of nature. The possibility of extending this definition to intersystemic and extrasystemic laws of any finite rank can be easily shown. It is important that these other groups of laws of nature can be exactly defined because the constitutive extrasystemic laws of rank 2 are of decisive importance to physical science. This follows from examples of laws of nature used in the opening part of this discussion: both the special and the general principles of relativity and Wigner's "superselection rules" are extrasystemic and of rank 2. The fact that, in some cases, a group of laws is

33 B. Kursunoglu, *Modern Quantum Theory* (San Francisco: W. H. Freeman Co., 1962).

promoted to the rank of "principle" is completely irrelevant. The second Principle of Thermodynamics can be expressed in the object language of this theory and is, nevertheless, often dignified with the title of "principle."

The following conditions must obviously be fulfilled by any adequate definition of intrasystemic laws of nature of rank 1:

(1) An adequate definition of a law of nature must be effective or "operational."[34] This means that it should be possible in principle to find out whether any preassigned statement in the object language of a physical theory expresses a law of nature. I am aware of the unsurpassable difficulties raised by Bridgman's requirement of operational definability of all concepts involved in physical laws. However, I am not prepared to reject the Theory of General Relativity on the alleged ground that the operationally indefinable concept of a law of nature is involved in the theory.[35] The only point I would like to make at this juncture is that, according to conclusions reached in this study, the concept of a law of nature is operationally definable. Bridgman may have overlooked this decisive circumstance because he was unaware of the existence and significance of laws of nature the logical rank of which $\neq$ 1.

(2) A satisfactory definition of a law of nature should enable us to tell a universal, physical law from a particular factlike statement which ascribes to a specific physical object X a specific property which X actually possesses, or, if a mathematical formalism involving spatio-temporal variables is used to make such a factlike statement, then we may expect that a definition of a universal law should not apply to statements in which any particular value of these coordinates occurs. In other words, laws of nature do not deal with any local events or with any individual, physical systems. We shall call this type of universality which any law of nature should exemplify *conceptual universality*.

The distinction between universal laws and particular facts is often construed by the mathematical physicist in terms of classical,

---

[34] P. W. Bridgman, *The Logic of Modern Physics* (New York: Macmillan, 1927).

[35] P. W. Bridgman, "Einstein's Theories and the Operational Point of View," in *Albert Einstein: Philosopher-Scientist*, ed. P. A. Schilpp (New York: Tudor Publishing Co., 1951) [Now published in La Salle, Illinois by Open Court Publishing Co.].

mathematical analysis, particularly in terms of the theory of partial differential equations. I take it that, in spite of the present shift towards functional analysis and the theory of group representations associated with an unexpected emphasis on topology (or, as Poincaré may have put it, *analysis situs*), there is still nothing wrong with referring to partial differential equations[36] in a consideration aiming at drawing a line between universal laws of nature and particular facts. The most typical laws of classical physics were expressed ordinarily in partial differential equations and involved physically significant magnitudes defined over the spatio-temporal continuum. On the other hand, a solution to the differential equations interpreted as laws of nature was a function determining a spatio-temporal distribution of the relevant quantities over this continuum. It was obvious that, in order to obtain any particular solution, i.e., to compute the actual spatio-temporal distribution of the physical quantities, supplementary data were required, in addition to the physical laws. These data were determined by initial or boundary conditions,[37] depending on the structure of the laws of nature replaced by so-called asymptotic conditions, or conditions prevailing "at infinity." This terminology, borrowed from the mathematical theory of differential equations, is very useful in attempts at defining laws of nature and preventing them from coinciding with particular physical facts.

At this juncture, it seems advisable to recall the studies of physical time made by O. Costa de Beauregard.[38] He reaches the conclusions that the existence of irreversible physical processes and the asymmetry of time deducible therefrom is not derivable from either classical or statistical thermodynamics. According to this investigator, the existence of irreversible processes is due entirely to the asymptotic boundary conditions; this idea is supported by presently available, observational data. However, regardless of what boundary conditions are known to prevail, their nature could not possibly affect the sum total of laws of nature as construed in this study. If the laws of nature are invariant under time reversal

---

[36] R. Courant and D. Hilbert, *Methods of Mathematical Physics* Vol. II (New York: John Wiley & Sons, 1953).

[37] *Ibid.*

[38] O. Costa de Beauregard, cf. Ref. 18.

(which is a sufficient, but not a necessary condition[39] for the irreversibility of physical processes, or the asymmetry of time) , then this type of invariance could not possibly be affected by the boundary conditions used to determine individual solutions of the equations which express the physical laws.

(3) A statement S possessing the conceptual universality of a law of nature would nevertheless not be classified under the heading of such a law if S happens to be false. Nor would a particular statement S, ascribing a specific property to a specific individual or the occurrence of a local event, be considered as the statement of a physical fact, unless S were true. We thus reach the conclusion that all laws of nature are lawlike statements but that the converse need not be true. Similarly, all statements of physical fact are factlike statements, but the truth-value of the relevant particular statement decides whether or not it states a fact.

The concept of "truth" used here to tell laws and facts from merely "factlike" and "lawlike" statements has no metaphysical implications or antiscientific connotations. The truth of logical and mathematical statements couched in a formalized language has been rigorously and noncontroversially defined by A. Tarski[40] in the branch of mathematical logic initially called "semantics," and presently usually referred to as "model-theory." This area of logical research is appreciably more advanced and better entrenched than the elementary logical theories usually referred to as the "propositional" and the "first-order logical function calculus."[41] On closer analysis, Tarski's model-theoretical definition of truth calls for far-reaching modifications if an extension of this definition to empirical statements is required. These statements need not be couched in a formalized language the basic terms of which are intuitively meaningful and whose logic coincides with the classical, two-valued logic governed by the Principle of Excluded Middle. In a treatise on the logical structure and potentialities of science,[42] I have explored the problems inherent in this extension of the model-theoretical defini-

---

39 Cf. end of Section III above.

40 Cf. Tarski, Ref. 27.

41 D. Hilbert and W. Ackermann, *Principles of Mathematical Logic* (Bronx, New York: Chelsea Publishing Co., 1950) .

42 Cf. Ref. 14, pp. 243-80.

tion of truth to the area of empirical science in considerable detail. The results obtained in this treatise show both the feasibility of this extension and the profound changes it requires. However, the redefined concept of truth is as free of speculative, nonscientific connotations as was the concept of mathematical truth related to formalized languages which Tarski's work made first accessible to exact, scientific procedures.

I shall now construct, in a few consecutive stages, a definition of laws of nature which fulfills the aforementioned conditions. Since the attempted definition will only apply to intrasystemic laws of nature of rank 1, we can assume that all the laws of nature under consideration can be expressed in terms of operator-valued[43] or c-number[44] valued functions defined over a set of space-time coordinates associated with an appropriate frame of reference. The logical formative rules show that any statement of this type can be built up gradually by performing the standard operations of negation, disjunction, quantification and substitution on the set of so-called "atomic" or "elementary" lawlike formulae of the relevant object language. An "atomic" law like formula of degree N is a function of N space-time coordinates of the type described. In the sequel, we shall realize the extreme importance of the fact that all presently available physical theories are expressible in terms of a small number of such functions.

Given these considerations, a definition of a law of nature can be obtained provided the following questions are answered: (1) Since the conventional view that every law of nature deals only with measurable quantities is abandoned in the proposed definition, how can the essential requirement of measurability be replaced? (2) How can we define the few functions or relations involved in laws of nature? (3) How can we eliminate the infinite class of statements which meet all our requirements but obviously have no law-status?

(1) The requirement that physical quantities be measurable in principle is of fundamental philosophical and physical importance. This requirement is also conspicuously vague and logically hazy. We must not forget, however, that A. Einstein was motivated, in his

---

43 Cf. Ref. 28, pp. 254 ff.
44 Cf. Ref. 28.

discovery of the theory of special relativity, by the requirement that
the time-interval separating two noncoinciding events be measurable in principle,[45] rather than by the desperate need to rid physical
science of the predicament inflicted on physics by Michelson's measurement of the speed of light. Similarly, the Equivalence Principle
of General Relativity rests basically on the impossibility of measuring the local discrepancy between a gravitational field and a field
conjured up by the choice of an appropriate coordinate system.
Heisenberg's[46] quantum mechanics seems to have been invented by
him mainly because of his belief that the quantities inherent in
Bohr's theory but not measurable in principle should be eliminated. There is no need to extend the list of significant cases where
the emergence of a physical theory was due to an application of the
measurability requirement. This holds not only of physics proper,
but of the philosophy of physics as well. Wigner's[47] interpretation
of a law of nature suggested in 1965 and mentioned several times
already in other contexts is another case in point.

If we were able to afford a purely mathematical way of thinking
in this investigation (which we obviously are not), then the requirement of measurability in principle could be easily made applicable to all laws of nature under consideration. It would suffice to
use von Neumann's[48] mapping of statements into projective operators. The validity of a law of nature would then consist in the
circumstance that the projective operator associated with the statement of the law has the eigenvalue 1. This reply, however, although
by no means refutable, would hardly be illuminating. To my mind,
it is more conducive to philosophical clarity and scientific validity,
if we do not take for granted the meaning of the measurability in
principle of a physical quantity but rather define the requirement
of measurability in terms of the empirical verifiability in principle
of any statement which ascribes any value to this quantity, for a
given physical system and a given frame of reference.

---

45 A. Einstein, *Annalen der Physik*, 17 (1905).

46 W. Heisenberg, *The Physical Principles of Quantum Mechanics* (Chicago:
University of Chicago Press, 1930).

47 E. P. Wigner, cf. Ref. 29, pp. 598 ff.

48 J. von Neumann, *Die mathematischen Grundlagen der Quantenmechanik*
(Berlin: J. Springer Verlag, 1932) pp. 130 ff.

In view of shortage of space, I shall refer the reader to an earlier, detailed investigation of mine[49] centering about the requirement of empirical verifiability in principle of statements transcending pure logic and mathematics but deserving scientific status. The main conclusions of this investigation, based on an accurate analysis and a precise definition of the concept of empirical verifiability in principle, can be stated as follows: If the concept of a true empirical statement is defined in the aforementioned way, then one can prove that statements which are empirically unverifiable in principle have no definite truth value, i.e., they are neither true nor false. Such statements, however, are by no means meaningless. They actually discharge an indispensable function in the most fundamental empirical theories. My conclusion is not ultrapositivistic, it does not classify empirically unverifiable statements as meaningless but restricts severely their cognitive import.

A quantity is measurable in principle if the statement ascribing any particular value to this quantity is verifiable in principle. Consequently, instead of requiring that laws of nature deal only with quantities that are measurable in principle, we shall postulate that these laws should be expressible by statements which are empirically verifiable in principle.

(2) The functions or relations admissible in laws of nature can be defined in exactly the same way as the statements pertaining to pure logic are defined. The basic vocabulary of logic which consists of all undefined logical constants was known to be very rudimentary ever since Frege[50] published his pioneering work. It can be reduced to two words, if technicalities are not avoided. The possibility of reducing the basic logical vocabulary to the three logical constants which designate the negation of a statement, the disjunction of two statements and the existential quantification of a statement was established in the classical *magnum opus* of Russell and Whitehead.[51] Consequently, a logical statement can simply be defined as any statement involving at most the above three logical constants in addition to variables and derivable from a few postulates which can also be defined by explicit enumeration.

---

49 Cf. Ref. 14, pp. 330 ff.

50 G. Frege, *Begriffschrift* (Nebert, 1879).

51 A. N. Whitehead and B. W. Russell, *Principia Mathematica*, I-III (New York: Cambridge University Press, 1910-12).

It is a most remarkable feature of physical science that its basic physical vocabulary can be reduced to a very short list. The list may include the values of the electric, the magnetic, the gravitational field strength at a particular world point, and a few more similar quantities. Hence, the concept of basic physical predicate can be defined by explicit enumeration. Eventually, the set of all statements qualifying as laws of nature of logical rank 1 can be determined in terms of the basic vocabulary and the admissible logical operations enabling the scientist to formulate meaningful statements the extralogical vocabulary of which coincides with the aforementioned basic, physical vocabulary.

(3) In this investigation, it is of particular importance to answer this third question adequately. It is obvious that even if the previous conditions which a statement has to meet in order to be a law of nature are fulfilled, some statements which obviously realize these conditions are not classifiable as laws of nature. Thus, suppose that a law of nature determines the dependency of the electromagnetic potentials upon the spatio-temporal distribution of the density of electric charges and electric current. It goes without saying that this statement would not be a law of nature if it were not reliably known to be true. I have already mentioned that the concept of truth as applied to empirical statements can be defined in such a way as to agree with the actual meaning of truth and not to entail any speculative consequences. Being reliably known to be true is a condition which has to be fulfilled by statements expressing laws of nature. Moverover, it is always possible in principle to find out whether or not this condition is fulfilled by any particular statement which also satisfies the requirement of being expressible in physical terms. The meaning of this requirement was clarified in answer #2. But it certainly is not the case that every statement expressible in physical terms and reliably known to be true is a law of nature. The point is that a physically meaningful and true statement S may deal with a particular space-time region, or a particular spatio-temporal point. Moreover, S can correctly evaluate the distribution of particular values of a specific, physical quantity over the spatio-temporal set under consideration. Under these circumstances, the status of a law of nature would certainly nevertheless be denied to S. The physicist would classify S under the category of boundary conditions or initial conditions, depending upon

the type of laws of nature with which the statement S is associated. In this context, we must also bear in mind the so-called asymptotic boundary conditions which are exemplified in the S-Matrix[52] approach to quantum physical collision problems. In any event, statements which can be dealt with as initial conditions, or boundary conditions (possibly of the asymptotic variety) are never considered as laws of nature. I shall not go into the technicalities required to rule out initial and boundary conditions statements.

The particular relevance to problems of quantum physical time of the conditions to be met in connection with question #3 can be shown very simply. The invariance under the expected Lorentz transformation of time and space coordinates and under time-reversal cannot be anticipated and actually does not obtain if statements dealing with initial conditions or boundary conditions are concerned. This point explains why the views I here present do not conflict with the views on time-asymmetry often ascribed to physical statements which do not fulfill our conditions of spatio-temporal universality. Thus, the Bondi-Gold[53,54] cosmological views on the expanding universe and the creation of matter, if true, would not provide a counterexample refuting the definition of law of nature proposed in this paper. I am using the "steady state theory" as an example illustrating the meaning of a law of nature. It does not matter that, in the meantime, this cosmological theory was refuted by observational results.

III. *The Symmetry of Quantum Time*

I now proceed with the analysis of a single, quantum physical discovery concerning problems of time. The discovery is ultimately traceable to J. Schwinger and frequently referred to in the physical code as the TCP Theorem. Significant extensions of the Schwinger insight are primarily due to Pauli and Lüders. The TCP Theorem has a valid epistemological and ontological version and can be

52 N. N. Bogoljubov and D. V. Shirkow, *Introduction to the Theory of Quantized Fields* (New York: John Wiley & Sons, 1959), pp. 192 ff.

53 H. Bondi, Cosmology (2d ed.; Cambridge: Cambridge University Press, 1961).

54 T. Gold, "The Arrow of Time," in *La structure et l'évolution de l'univers* (Brussels: R. Stoops, 1958).

formulated, in the more familiar, epistemological version as follows: "Every relativistic quantum field theory is invariant under the consecutive reversals of time, charge, and space." To determine the effect of this quantum physical discovery on the symmetry of time, I shall, first, indicate the way in which the above three reversal-operators are construed and show then, that the substitution of a relativistic interpretation proves, for the conventional interpretation of TCP, unmistakably that the TCP Theorem establishes actually the symmetry of quantum physical time and the invariance of laws of nature under an adequately defined time-reversal.

If we used only the conventional, physical terminology, we would have to put on a par the relativity of time and space. This terminology, however, is extremely misleading. It makes no sense to discuss either relativistic time, or relativistic space separately because none of them has an invariant, autonomous meaning. It stands to reason that only the status of a physical entity corresponding to an inherently meaningful idea can be explored. The adequate concept in this investigation is therefore Einstein's "proper time," rather than a "temporal" component of the invariant, relativistic space-time continuum, or, in Minkowski's[55] parlance, a temporal component of the "world." The interval of proper time which separates two world-points or events is invariant, and, obviously, a "Minkowski Interval." So are the "space-like" intervals Minkowski intervals. Since the four-dimensional world-continuum cannot be invariantly split into a one-dimensional time and three-dimensional space, it is necessary to view this continuum as a four-dimensional field metricized in terms of Minkowski intervals.

The consideration I am presently submitting is acceptable, of course, both in quantal and prequantal relativity. We shall see, however, that the logical structure of relativistic quantum field theories provides a new and cogent argument in support of the fieldlike structure of the world-continuum. In other words, this continuum must be considered as the "universe of discourse"[56] of the pseudo-Euclidean world, if this metamathematical way of speaking may be used in this context. This means that whatever exists in

---

[55] H. Minkowski, "Space and Time," in *The Principle of Relativity* (London: Methuen & Co. Ltd., 1923).

[56] Cf. Ref. 14, pp. 92 ff.

the physical universe is either part of this continuum or consists in a relation of a finite degree and finite level based on this "world." (The "degree" of a relation is the number of entities among which it obtains. The rank of the relation that is based on the world cannot be explained here without too many technicalities.[57] In any case, the only "individuals"[58] which can be referred to, or described, or talked about in a physically interpreted formalism of a relativistic quantum field theory are the events or points of the "world.")

Before embarking on an analysis of the TCP discovery, I have to comment shortly on how the three "reversals" involved in "TCP" are construed. In most typical cases, time reversal is self-explanatory. However, if a law of nature involves Dirac spinors, e.g., then this reversal must be adequately adjusted and does not consist in the substitution of $-t$ for t. In such cases, the reversal of time is called the "Wigner time reversal."[59] One of its essential properties, discovered by Wigner at an early phase of quantum physics, is that the reversal-operator is antiunitary. The reversal of space is always a unitary operator and consists, basically, in a reorientation of three mutually perpendicular directions of space. The reversible, generalized charge is a natural quantum generalization of an electric charge, the atomicity of which was discovered by Millikan. The incessantly growing number of quanta of matter or elementary particles explains the need for a generalized charge: (1) the *leptonic* charge is an additive quantum number characterizing leptons, i.e., quanta with a rest-mass inferior to that of nucleons. (2) *Baryonic* charge is an additive quantum number characterizing baryons, i.e., quanta with a rest-mass which exceeds or equals that of a nucleon. When a generalized charge of a physical system is involved, then at least one of the above types of charges is relevant. The reversal of a generalized charge is usually referred to as *charge-conjugation* or as *particle-antiparticle* substitution. The generalization of charge is warranted by the multiplicity of elementary par-

---

57 R. Carnap, *Symbolic Logic* (New York: Dover, 1958) , pp. 65 ff.

58 W. V. O. Quine, Ref. 19, p. 157 ff.

59 E. P. Wigner, *Göttingen Nachrichten,* Mathematisch-Physikalische Klasse **31** (1932) , p. 546.

ticles. The physical meaningfulness of generalized charges is sub-
stantiated by the possibility in principle of determining their values.

The specific relevance of the TCP discovery to physical time and
to the new role which it is now known to play in the universe, in
view of the sum total of quantum physical findings, is not apparent
from the conventional formulation of the theorem. We do notice
that the reversal of time is involved, but only in the presence of a
spatial inversion and of a charge conjugation or reversal. Moreover,
we all vividly remember the recent discovery that at least one of
these reversals, viz. the reversal of spatial dimensions,[60] does not
give rise to any "Principle of Invariance of Laws of Nature under
Space Reversal." The violations of this particular invariance principle rank with the most significant experimental and theoretical
discoveries of "Quantum Field Theory." I have already mentioned
that, on closer examination, the entire area of empirical science at
its present stage is actually involved.

At this juncture, the question arises naturally as to the reasons
why charge conjugation or charge reversal should be relevant in a
fundamental law dealing with the undirectionality of the world.
The answer is suggested by the study of other fundamental theories,
e.g., classical, Maxwellian electrodynamics which is usually regarded
as invariant under time-reversal and incapable of providing time
with an "arrow." The point is that, in order to show that the
equations which constitute Maxwell's theory remain unchanged
under time reversal, it is not sufficient to change the sign of all the
time variables occurring in these equations. In addition, we must
also reverse the magnetic field vectors. The need for reversing the
magnetic vector is easily understood since, according to the classical
doctrine, there is no "true magnetism,"[60a] and the magnetic field
vectors owe their existence to nonrectilinear motions of electric
charges. Since a reversal of time would automatically reverse these
motions responsible for the magnitude and direction of magnetic
vectors, it comes as no surprise that a time reversal should be
accompanied by a reversal of the magnetic field.

It is obviously necessary to consider the assumptions made by

---

[60] T. D. Lee and C. N. Yang, *Physical Review*, 104 (1956).

[60a] For an alternative approach suggested by J. Schwinger, cf. *Science*, 165, No. 3895 (1969).

Wolfgang Pauli[61] in his proofs of the TCP Theorem in order properly to evaluate the validity and implications of this theorem. Pauli made use of the following hypotheses in his proof of the TCP Theorem:

(1) The field-equations expressing the laws of nature which govern the interdependence of the main physical quantities are of a local type.

(2) These equations constitute the *Euler-Lagrange* equations of a Least Action Principle with the understanding that the action to be minimized is a time integral of a Lagrangian and that the latter is invariant under a Lorentz transformation with a determinant $=1$.

(3) The field-operators at points of space-time separated by a spacelike interval either commute or anticommute depending on whether their *spin* is integral or half-integral.

(4) Particles the statistics of which is of the *Bose-Einstein* type are possessed of quantities represented by operators which commute with each other. Particles whose statistics are of the *Dirac-Fermi* type, if kinematically independent, *anticommute*.

(5) Every product of field-operators of a Boson-field is completely symmetrized in contrast to such products related to Dirac-Fermi fields. The latter are completely antisymmetrized.

There is no point in outlining here the proof which Pauli has constructed on this postulational basis. Nor shall we indulge in summarizing the technically different proof given by Lüders[62] whose demonstration is not derived from an operator algebra. I shall similarly skip an ingenious proof which Jost[63] derived from Wightman's axiomatization of quantum field theory. The main point of interest in these investigations is the assumption of the invariance under a transformation of the Lorentz group for the quantum field theory under consideration. The view of time and space inherent in Einstein's Special Theory of Relativity is, to my mind, essential to the understanding of the TCP Theorem and to the realization of its significance.

---

[61] W. Pauli, cf., Ref. 2.

[62] G. Lüders, *Annalen der Physik*, 21 (1957).

[63] R. Jost, *Helvetiae Physica Acta*, 30 (1957).

I shall now evaluate the significance of the assumptions made in Pauli's proof which were readjusted in the derivations of Jost and Lüders. The two questions that matter in this investigation deal with how much they have proved and what additional assumptions they had to make in order to obtain a cogent proof. In the opening section, I have listed two informally worded problems of quantum time: The physical symmetry of time including both the time reversal invariance of physical laws of nature and the alleged occurrence of irreversible, physical processes. The main objective of the present section is to ascertain to what degree the TCP Theorem affects the problem of temporal symmetry. Perhaps, prior to embarking upon an analysis of the relevance of the TCP Theorem to temporal symmetry, it may be advisable to summarize in a single paragraph the remarkable ontological and metaphysical implications of temporal symmetry.

If all natural laws are time reversal invariant and no irreversible processes occur in the physical universe then there is no inherent, intrinsically meaningful difference between past and future—just as there is no such difference between "to the left of" and "to the right of." If this is actually the case, then all mankind's major religions which preach a creation of the universe (by a supernatural agency) and imply, accordingly, a differentiation between the past and the future, i.e., an intrinsic difference between both, would have to make an appropriate readjustment of man's major religious and "creationist" creeds and the scientific findings. This readjustment need not raise unsurmountable difficulties. The emergence of Darwin's theory of biological evolution raised similar problems from a metaphysical point of view. Bergson's work on creative evolution showed that the issue is not unsolvable. The specific metaphysical issue related to time and mentioned at the beginning of this paragraph will not be treated in this investigation. I have pointed out this issue only for illustrating the ontological implications of the quantum theorem concerning the invariance of the physical laws of nature under the three consecutive transformations of time reversal —matter, antimatter substitution and space reversal.

In view of the unique significance of the quantum physical discovery traceable to J. Schwinger, I shall also indicate some physical implications of this discovery which show clearly the ontological relevance of the TCP Theorem despite its conventional, epistemological wording.

Let us consider the following physical processes: (1) A quantum-system consisting of N elementary particles E_i (i = 1,2,,,,N) is characterized at time t_1 by the positional coordinates (X_i,Y_i,Z_i) of all the relevant particles, the components of linear momentum (L_i,M_i,N_i) and the spin components (S_i,T_i,U_i). At another time-instant t_2, the positional and linear momentum coordinates are similarly designated by X'_i etc., L'_i etc. and S'_i etc. (2) The quantum system qs differs from $\overline{qs}$ in the following way: (a) The particles of qs are replaced with the corresponding antiparticles in qs. (b) The time-instants t_1,t_2 are interchanged in the system $\overline{qs}$. (c) The positional and spin-coordinates of qs are reversed in $\overline{qs}$. Then a physically significant consequence of the TCP Theorem states that the frequency of the transitions effected by the first quantum system during the time interval separating t_1 and t_2 is exactly equal to the frequency of the transitions which the second quantum system underwent during the same time-interval, with the understanding that the limiting instants of the first interval have been reversed.

This pedantic formulation of an important, physical consequence of the TCP Theorem was necessary to show the precise conditions of its applicability. I am skipping some other technicalities, e.g., the reason why the components of the linear momentum of the first quantum system must not be reversed when the second system is taken into consideration. The physical law derivable from the TCP Theorem shows, in any event, that the *ontological* problem of time symmetry is also involved in this quantum physical finding because the latter deals with relative frequencies of physical processes rather than with invariance properties of physical laws. In terms of the classification of laws of nature based on their logical rank, which I have introduced in Section II, the above consequence of the TCP Theoerm is a law of nature of rank 1.

Since I have embarked on a few significant technicalities, I have to point out that the concept of time reversal, to which I have frequently resorted, must not be oversimplified. Thus, logically different ways of reversing time are required depending upon whether the elementary particles involved in the relevant physical processes fall into the category of bosons or fermions. This dichotomy determines the social behavior or the average behavior of elementary particles, according to another, fundamental discovery made by Pauli. Consequently the aforementioned, physical consequence of the symmetry of time also describes the statistical behavior of quantum systems. The difference between the two types of Wigner time reversal which are associated with the two types of particles is far from trivial, as shown by other significant but not unobjectionable definitions of time reversal, e.g., the Racah time reversal.[64]

Other aspects of the ontological symmetry of quantum physical time may be mentioned although, on a closer analysis, they are not unrelated to the proper time symmetry inherent in the TCP Theorem. Stueckelberg[65] and Feynman[66] discovered that the relativistic quantum mechanics of positrons can be obtained by applying Wigner's time reversal operator to Dirac's spinorial mechanics of electrons. Their result provides perhaps the most impressive insight into the physical symmetry of time. Furthermore, the causal interaction of systems of elementary particles is entirely described by the so-called "Interaction Hamiltonian."[67] This also shows the physical significance of time symmetry. Another illustrative example is apparent from a basic quantum theoretical approach to all conceivable, causal interactions of systems of elementary particles: the so-called "S-matrix theory." This quasiphenomenological theory, traceable to Heisenberg, correlates causally the asymptotically initial and asymptotically final quantum-state of the relevant physical system of elementary particles. One of the fundamental, physical consequences of the S-matrix theory is the fact that no physical change would occur if the initial and final, asymptotic states were interchanged. The time symmetry of quantum processes described in the S-matrix theory in terms of asymptotic initial and final states of a physical system can be significantly extended according to a result obtained by A. S. Wightman in 1956[68] on the basis of

64 G. Racah, *Nuovo Cimento*, 14 (1937).

65 E. C. Stueckelberg, *Helvetiae Physica Acta*, 14 (1941).

66 R. P. Feynman, "Theory of Positrons," *Physical Review*, 76 (1941).

67 Cf. Ref. 52, pp. 83 ff.

68 G. Feinberg, *Physical Review*, 108 (1957).

previous results of V. Bargmann and Hall.[69] The Wightman result establishes the physical possibility of introducing "interpolating fields" into the S-matrix theory without the need of any additional assumptions and, even in association with a substantial extension of the general quantum field theory. The ontological symmetry of time follows logically from this valuable version of quantum theory. Moreover, in view of a few technicalities which cannot be described in this context, the aforementioned Wigner time reversal can be replaced with the classical time reversal, i.e., the interchanging of the past and the future.

In connection with my discussion of the epistemological and ontological version of the problem of time symmetry, a few comments on these two types of problems concerning physical time are called for. From a historical point of view, the remarkable phenomenon is the conspicuous, contemporary shift in the scientist's interest, viz. from the ontology towards the epistemology of time. During the nineteenth century, when phenomenological thermodynamics and its statistical extension became overnight the center of scientific endeavor, only ontological time issues were explored. Irreversible physical processes, entropic changes and related aspects of the universe were on the minds of the men who created the new physical theory: Carnot, Gibbs, Maxwell, Boltzman and Clausius were fascinated only with the ontological enigma of time. To a considerable extent, this also holds of our century, as shown by the interest which Einstein, Gödel[70] and von Neumann[71] took in the question of "time's arrow." As a matter of fact, the first, significant quantum physical contribution to the problem of time symmetry was made by Einstein as early as 1910.[72] He pointed out that the ostensible irreversibility of a fundamental, physical process, viz. the spherical propagation of the light emitted by a pointlike source, vanishes, provided that the wave-model of light be replaced with the particle model. Von Neumann's interest in the ontological symmetry of time will be shown in Section IV. This holds also of

---

69 P. Roman, *Advanced Quantum Theory* (Reading, Mass.: Addison-Wesley, 1965) , pp. 314 ff.

70 K. Gödel, Ref. 18.

71 Cf. Section IV above.

72 A. Einstein, *Physikalische Zeitschrift,* 10 (1910) .

other investigators who attempted to extend the laws of entropy to quantum physical areas, e.g., Szilard,[73] Pauli,[74] Fierz,[75] and Ludwig.[76]

Other attempts at exploring the "direction of time" made during the last few decades should be mentioned. Schrödinger tried, unsuccessfully, to derive the irreversibility of entropic processes from a modified version of the second law of thermodynamics.[77] Szilard made a most significant contribution to the emergence of information theory because of the interest he took in Maxwell's thermodynamical demon. Feynman[78] suggested that the asymmetrical aspects of time in the macrophysical area may be entirely due to the extraordinary order of magnitude of sets of elementary particles involved in the transition from micro- to macrophysics. It would be too easy to extend this list in order to show that the fascination with the ontology of time has never subsided.

Yet, there is no doubt as to the present emphasis on the epistemological issue of time invariance. In a number of important treatises on quantum physics, the question of whether the laws of nature are invariant under time reversal is discussed in detail, although terms like "entropy" or "irreversibility" never occur. The present interest in time invariance of laws of nature is probably traceable to Wigner's work. I would like to make two points in this context: (1) The principle of invariance of laws of nature under time reversal is itself a fundamental law. However, the logical rank of this principle is 2, whereas all laws of nature dealt with in the principle are intrasystemic and of rank 1. (2) On closer examination, the logical interrelatedness of the ontological time symmetry and the epistemological time covariance turns out to be an asymmetrical logical entailment. The validity of the epistemological principle is a sufficient but not a necessary condition for the validity of its ontological counterpart.

The important conclusion which can be derived from all avail-

73 L. Szillard, *Zeitschrift für Physik* (1932).

74 W. Pauli, cf. Ref. 5.

75 Cf. Ref. 6.

76 G. Ludwig, Ref. 7.

77 E. Schrödinger, *Proceedings of the Royal Irish Academy*, 53 (1950).

78 Cf. Ref. 15, pp. 235 ff.

able proofs of the TCP Theorem is that the validity of this fundamental result is conditional on the validity of the relativistic view of
space and time. We shall see that, from the relativistic point of
view, any statement made about either a spatial or a temporal
component of Minkowski's world is completely vacuous. The only
physically invariant and inherently meaningful aspect of the four-
dimensional world-continuum is the proper-time which elapsed between any two physical events, or, more generally, of their Minkow-
ski-interval. Consequently, the fact that all relativistic quantum
theories are invariant under the TCP transformation or that the
four-dimensional continuum is symmetrical actually implies an attribute of proper-time. Granted, that the symmetry or isotropy of
proper-time is interrelated with the symmetry of the generalized
charges. The interconnection of time reversal and magnetic field
reversal was, however, already inherent in the prequantum physical
world picture as determined by Maxwell's electrodynamics and
Newton's or Einstein's mechanics. Both connections illustrate the
interrelation of Minkowski's world with matter.

The main point of the TCP Theorem, however, is the claim of
the invariance of all laws of nature under the three consecutive
reversals of time, generalized charge and three perpendicular spatial
directions. The TCP Theorem does not imply that all laws of
nature of rank 1 remain valid when only one of three magnitudes is
reversed. In particular, if only the direction of time is reversed,
while the spatial directions and the generalized charge remain
unchanged, then it is consistent with the TCP Theorem that the
laws of nature governing systems subjected to this single change
would not remain invariant. Granted, this is reasonably established
today only for the reversal of spatial directions ("parity violations"). The existence of such laws was refuted in my earlier papers
of basic physical theories. The compatibility of time-reversal invariance with the behavior of electrically neutral K-mesons can also be
shown, and is dealt with in my forthcoming work on *The Problem
of Time*. The references to significant papers are included in this
volume. So is an acknowledgement of relevant discussions with Dr.
R. C. Sachs, of the University of Chicago. In any event, laws
violating the time-invariance may come to be established, sooner or
later.

There are some doubts, however, whether such laws would be *cosmic*, i.e., describe processes which go on always and everywhere. Should laws violating time-invariance fail to be cosmic then we could hardly admit that they provide physical time with an arrow, or a privileged direction. In its entirety, physical time would still have to be symmetrical. The humanly fundamental difference between past and future should then be attributed to man's special orientation in physical time, rather than to time-proper. I cannot afford to discuss this issue in this paper, for lack of space. However, even if laws violating the invariance of time were eventually discovered, the TCP Theorem would essentially restrict their bearing on the nature of time and, in particular, on the intrinsic difference between past and future.

The point I would like to make at this juncture is that, in view of the TCP Theorem, no sequence of events could be accounted for in terms of time's arrow. Suppose that we follow the fate of a man who keeps getting older and eventually passes away. This man would remember, at each moment of his conscious life, a considerable part of his past experiences but he would have no comparable knowledge of the future ahead of him. The conventional view of time's arrow, shared by men like Eddington and von Neumann, implies an intrinsic difference between past and future, derivable from the laws of entropic change. It emphasizes that the succession constituting the life of the man is not an intrinsically meaningful entity since it depends on our choice of spatio-temporal coordinates used for a description of this life. The description would become different if we chose, instead, a system of coordinates reversed with regard to both time and space. Moreover, apart from this extrinsic system, the life of the man would also crucially depend upon the particles making up his organism. If we managed to replace these particles with their respective antiparticles while also replacing his coordinates with a reversed set of four coordinates, we would notice that the life of the second man described with the second set of coordinates would be governed by the same set of physical laws to which the first man was subject. Hence, we cannot make a privileged direction of time responsible for the man's life and death. Only in conjunction with the directions of space and the generalized charge can the direction of time run the man's life and death. There is no intrinsic difference between past and future. Time, space and generalized charge are, all of them reversible.

The bearing of the TCP discovery upon our views of the nature of physical time may therefore be described as follows: The invariance of laws of nature under time-reversal has not been refuted but it has been deprived of an autonomous significance. Only in conjunction with the reversal of space and charge, can the reversal of time be viewed at present as affecting the history of the cosmos and the laws governing this cosmos. A reversal of time can always be compensated by a simultaneous reversal of charge and space. In this triple correlation of time, space, and matter, there is nothing to justify that time has an arrow and that cosmic history and human life follow a direction indicated by this arrow. More specifically, we can make the assumption compatible with all available quantum theories that every particle collided with its antiparticle and the world consists of photons only. Such a world would be entirely governed by quantum electrodynamical laws. Quantum electrodynamics is known to be invariant under time-reversal.

IV. *The Problem of Reversibility in Quantum Theory*

A brief account of the relevance of investigations into quantum thermodynamics to the problem of time symmetry will now be presented. I realize that the relativistic reinterpretation of the TCP Theorem discussed in III should suffice to dispose of attempts made to derive the asymmetry of time from quantum physical extensions of thermodynamics. However, the significance of the presently neglected, ontological problem of time symmetry seems to justify an analysis of microreversibility: this is the only vital quantum physical issue which affects the status of time symmetry.[78a] I shall not discuss the interesting results obtained by G. Ludwig[79] and other physicists active in axiomatic quantum theory. Only von Neumann's[80] contribution to the problem of microreversibility will be

---

[78a] At this juncture, the behavior of electrically neutral K meson particles is not taken into consideration. This issue is analyzed in my work: *Time in a Quantized Universe*, to be published shortly by the American Association for the Philosophy of Science. For a recent discussion cf. R. C. Casella, *Physical Review Letters,* 21 (1968) and G. E. Harrison, P. G. H. Sandars, and S. J. Wright, *Physical Review Letters,* 22 (1969).

[79] G. Ludwig, Ref. 7.

[80] J. von Neumann, Ref. 4.

considered. To my mind, his ideas of time symmetry are untenable in view of the difficulties to be discussed in this Section. These ideas are nevertheless of vital interest to contemporary quantum physics. The remark which Wigner[81] made in 1963 about the need of "relearning von Neumann" is relevant here.

Von Neumann's stand on time symmetry is clearly expressed in his statement that Newtonian mechanics is unable to describe "one of the most essential and striking aspects of the real world: the fundamental difference between the directions of time called 'past' and 'future.'[82] He used two types of argument in support of microphysical irreversibility: (1) The possibility of deriving the existence of irreversible microprocesses from his quantum theory of measurement. (2) The possibility of deriving the monotonic increase of an entropy redefined in quantum physical terms from his quantum mechanical ergodic theorem. In this context he also made the claim that in contrast to Newtonian mechanics where additional "chaos-assumptions" were necessary for the derivation of the second law of thermodynamics, the quantum mechanical axioms provide for this derivation without any auxiliary hypothesis.

To realize the insufficiency of the argument derived from the theory of measurement and based, ultimately, on the quantum mechanical axioms it suffices to consider the logical structure of this set of axioms. In the main, the set includes: (a) Schrödinger's time dependent partial differential equation and (b) Born's axiom of statistical interpretability of nonrelativistic quantum mechanics. The invariance of Schrödinger's equation under time reversal was clearly shown by von Neumann. Born's interpretational axiom may be formulated as follows: Assume that (I) the quantity Q is represented, in the quantum mechanical formalism by a hypermaximal operator Q' defined over the relevant Hilbert space which can be realized by a set of square-integrable functions. (II) An ensemble E of quantum mechanical noninteracting systems is characterized by the quantum state corresponding to a single function Ψ in the Hilbert space. Then the expectation value of Q in E is equal to the scalar product of Ψ and of the Q' transform of Ψ. However, the

---

[81] E. P. Wigner, *The Problem of Measurement* (*American Journal of Physics,* 1963).

[82] Cf. Ref. 48, pp. 190-91.

expectation value of any quantum mechanical magnitude in an ensemble characterized by a single Ψ function is physically synonymous with the transition probability corresponding to the transformation of the initial state function Ψ of the ensemble E into any preassigned eigenfunction of Q'. Quantum mechanical transition probabilities are known to be invariant under time reversal. Consequently, von Neumann's second axiom is also time invariant. It is impossible to derive the existence of irreversible processes from any set of time invariant axioms.

Von Neumann's second attempt at deriving the asymmetry of time from quantum mechanical axioms is based on his ergodic theorems.[83,84] He first succeeded in proving the ergodic theorem[85] on the assumption that the relevant ensemble consists of physical systems obeying the laws of Newtonian mechanics. Subsequently, he extended the validity of the first ergodic theorem to ensembles of quantum mechanical systems. The quantum mechanical version of the ergodic theorem can be stated as follows: The phase averages of Gibbsian ensembles of quantum mechanical systems characterized by the same quantum state converge towards the same limit to which tend the entropies of the associated microcanonical ensemble. The equality of these two limits is clearly irrelevant to the irreversibility of entropic changes. Accordingly, no support for the asymmetry of time can be derived from the quantum mechanical ergodic theorem. The well-known fact that von Neumann overlooked a tacit "chaos-hypothesis" in his proof of the ergodic theorem is of no consequence in this discussion.

V. *The Relativity of Time in Quantum Physics*

The incessant efforts made during the last four decades by two generations of the world's physicists to build the relativity of time and space into quantum physics are the most dramatic scientific phenomenon of the century. The struggle began already in 1926 when Schrödinger tried unsuccessfully to attain this objective for

---

[83] J. von Neumann, *Proceedings of the National Academy of Sciences*, 18 (1952) , pp. 70-82.

[84] *Ibid.*, pp. 263-66.

[85] P. A. M. Dirac, *Proceedings of the Royal Society, London, A* 117, 118 (1928) .

his own quantum mechanics. Substantial successes were achieved when Dirac discovered his relativistic quantum mechanics and, a few decades later when Dyson,[86] Feynman,[87] Schwinger[88] and Tomonaga[89] succeeded in establishing a relativistic quantum electrodynamics. The experimental and theoretical implications of both relativistic advances belong to the most important experimental and theoretical achievements of twentieth century science, second only to what Einstein did. At present the struggle is far from over. The difficulties inherent in divergences, renormalization-procedures and related quantum physical problems[90] seem to indicate that the struggle will never end unless both the relativity of time and space and all the presently available quantum field theories undergo profound, unpredictable changes. It is not the object of this investigation to tell the epic story of these four decades of relativistic quantum physics nor speculatively to guess how it may eventually wind up. We want only to understand the nature of the difficulties which must be overcome, if the relativity of time and space has to be built into quantum physics and to evaluate the motivation and scope of the achievements obtained so far.

The motivation is very complex, of course. It is objective, to a large extent. The creation and annihilation of elementary particles, and the duality of matter and antimatter, which are both at the center of present physical research, seem to require that relativity of time and space be inherent in the relevant quantum theories. The heuristic value of the principle of relativity in the quantum physical area is also involved. Dirac's prediction of positrons, Pauli's discovery of the statistical implications of quantum spin, the uniquely accurate account of a spectroscopic effect[91] by Feynman[92] and Schwinger[93] give an idea of the methodological role of the

---

86 J. Dyson, in J. Schwinger (ed.), *Selected Papers on Quantum Electrodynamics* (New York: Dover, 1958).

87 S. Tomonaga, cf. Ref. 86.

88 R. P. Feynman, cf. Ref. 86.

89 J. Schwinger, cf. Ref. 86.

90 G. Källen, "Quantenelektrodynamik," *Handbuch der Physik*, 1 (1958).

91 W. E. Lamb and R. S. Retherford, *Physical Review*, 72 (1947).

92 R. P. Feynman, cf. Ref. 86.

93 J. Schwinger, cf. Ref. 86.

principle of relativity. The speeds of elementary particles either equal or close to the speed of light make the resort to a relativistic time and space imperative. There is no point in extending the long list of objective and subjective reasons for relativistic trends in quantum physics. To my mind, the soundness of the view that the real, physical time is governed by relativistic laws within and without quantum physics is the main reason for this scientific drive.

The nature of the difficulties raised by the relativity of time and space in physical contexts is not entirely understood at present. Some of the sources of these difficulties seem, however, quite obvious. The logically inadmissible infinitist divergences which beset relativistic quantum theories are partly inherited from nonrelativistic and nonquantal predecessors. For instance, the self-energy of the electron was as disturbing in Lorentz's theory of the electron as it is in quantum electrodynamics. Yet a closer analysis shows readily that the infinitist predicament in relativistic quantum physics cannot be entirely blamed on the gigantic mistakes committed by giants of the past. Moreover, the relativity of time and space in quantum physics is not physically meaningful, if Einstein's classical interpretation of relativity is used. I have discussed elsewhere the necessity of replacing his macrophysical, phenomenological views of relativity with a more comprehensive interpretation. An alternative solution to this problem will be outlined in the sequel. There is no point in stressing the importance of interpretive problems, in view of the role played by Born's statistical interpretation of quantum theories and by the "Copenhagen Interpretation" due to Bohr and Heisenberg. I do not think that identifying the relativity of time and space in quantum theories with the algebraic phenomenon of the insensitivity of these theories to the transformations of time and space which are associated with a change of the initial frame of reference is sufficient. The algebraic flexibility of these equations must be shown to reflect a significant property of the real, spatio-temporal structure of the physical universe if relativistic quantum physics is to be regarded as a body of reliable knowledge of the universe rather than as a predictive gadget now indispensable in many areas of human endeavor, including the area of nuclear engineering on which man's survival depends.

In addition to these infinitist and interpretive problems raised by the relativity of quantum physical time, a few other worries may be mentioned. As pointed out in another paper of mine,[94] the relativity of time and space raises completely new difficulties in quantum theories, in contrast to the easy manageability of classical theories. Thus, replacing Newton's mechanics and Maxwell's electrodynamics with the corresponding relativistic theories was easily and quickly effected by a few men of genius. This is now impossible because of the conceptual innovations of quantum physics. They can be illustrated by the novel idea of a quantum state associated with a "ray" in a Hilbert space or by the assignment to "hypermaximal" Hermitean operators of the role of mathematical models for fundamental physical quantities. There is no point in discussing whether Dirac was right when he felt that the idea of quantum state is the principal conceptual innovation of quantum theory.[95,96] Perhaps it is rather closer to truth if one here mostly emphasizes the significance of the operator concept in quantum physics (as Feynman[97] is apparently doing now). The essential thing, common to both Dirac's and Feynman's estimates of quantum physical, conceptual changes, is the central role now assigned to the mathematical theory called functional analysis, in contrast to classical mathematical analysis which entirely dominated prequantum physics and centered about the applications of the theory of systems of partial differential equations. The application of functional analysis complicated immediately the problem of time and space in quantum physics. When a physical theory has a formalism involving function spaces, the validity of the Lorentz transformation group for time and space coordinates is no more sufficient to guarantee the independence of a physical theory of such a split into time and space components. This is known in some cases and seems puzzling even in these cases, as Gel'fand[98] observed. Moreover, an infini-

---

94 H. Mehlberg, "Space, Time, Relativity," in *Proceedings of the 1964 International Congress for Logic, Methodology, and the Philosophy of Science* (Amsterdam: North Holland Publishing Co., 1966).

95 P. A. M. Dirac, *The Principles of Quantum Mechanics* (New York: Oxford University Press, 1930).

96 W. Heisenberg, "The Copenhagen Interpretation of Quantum Theory," in *Physics and Philosophy* (New York: Harper and Row, 1958).

97 R. P. Feynman, cf. Ref. 15, p. IX (1965).

98 M. Gel'fand, R. A. Minlos and Z. Ya. Shapiro, *Representations of the Rotation and Lorentz Groups* (New York: Pergamon Press, 1963).

tist complication of the problem of relativity of time and space is associated with the relativistic transformation of physically interpreted operators. According to results, first obtained by Wigner and now dominating quantum physics, these operators transform under usually infinite dimensional representations of the Poincaré group. Incidentally, the extensive and extremely effective use of the theory of group representations is not only a source of complications in relativistic quantum theories but also a second major innovation in the mathematical equipment of the physicist. I have in mind the trend started by Wigner[99] rather than the classical approach of H. Weyl. I would not like to comment on whether L. de Broglie did state that whatever can be done in quantum physics by applying group theory, can also be done without group theory.

Let us now stop deploring the quantum physical predicaments entailed by its new mathematical tools and discuss the difficulties of relativistic quantum physics. In a paper published two years ago, I have outlined some results concerning the necessity of completely reinterpreting the relativity of time in quantum physics and the possibility of axiomatizing relativistic world geometry in terms of the single quantum physical concept of collision connectibility.[100] This axiomatization established a new interpretation of the relativity of time, valid within and without quantum physics. In the present investigation I shall first indicate several deeper reasons for a new interpretation of the relativity of time and space. Then I shall outline an alternative, axiomatic system of relativistic space and time which may be better adjusted to these additional arguments in support of reinterpreting the relativistic theory of time in quantum physics.

The aforementioned argument showing the difficulties of Einstein's classical interpretation of the Special Principle of Relativity stressed his inherently phenomenological and macrophysical idea of an inertial frame of reference. The proposed axiomatic system established a new definition of an inertial frame of reference meeting the requirements of both quantum physical and other relativistic

[99] E. P. Wigner, *Group Theory* (New York: Academic Press, Inc.; 1959).

[100] H. Mehlberg, "Relativity and the Atom," in *Mind, Matter, and Method: Essays in Philosophy and Science in Honor of Herbert Feigl*, P. K. Feyerabend and G. Maxwell, (eds.) (Minneapolis: University of Minnesota Press, 1966).

theories. In the alternative axiomatic system, no redefined inertial frame of reference occurs.[101,102] The occupation number type of field formalisms, the theory of which made important progress since the publication of Dirac's radiation theory (perhaps mainly due to Fock's[103,104] theory of specifically interpreted Hilbert spaces) is as follows: The use of such formalisms was optional in several quantum theories but is inevitable in *quantum electrodynamics*. The logical necessity of this type of formalism in this relativistic quantum theory can be explicitly proved. One proof is outlined in a forthcoming paper of mine. At this juncture it may suffice to point out the following fact which shows that a fieldlike universe of discourse is inherent in quantum electrodynamics: The complete annihilation of particles with a nonvanishing rest-mass, associated with the corresponding increase of the amount of radiation energy in the universe, is compatible with the set of all quantum electrodynamical laws. During the time-interval characterized by a vanishing, cosmic rest-mass, the physical processes would still go on. Hence, since only the four-dimensional world-continuum would then exist in addition to the relations of various degree and rank over the continuum, we would have to reinterpret the relativity of time and space in such a way that it should involve no "material" (I mean, by "material": "possessing a nonvanishing rest-mass") entities. This quantum electrodynamical reason for reinterpreting the relativity of time and space seems more significant to me than the shortcomings of Einstein's macrophysical interpretation.

The necessity of selecting a field formalism of the occupation number variety for adequately expressing the fundamental laws of quantum electrodynamics will not be discussed in detail here, because of shortage of space. I shall only mention a result of an investigation which shows the epistemological necessity of the occupation number type of formalism. To my mind, a sound, new interpretation of a physical theory should fulfill the following conditions: (1) The interpretation must not be subjectivistic and make

---

[101] H. Mehlberg, "The Problem of Causality in an Indeterministic Science," forthcoming in *International Journal of Theoretical Physics* (1969).

[102] H. Mehlberg, the present essay.

[103] V. A. Fock, *Zeitschrift für Physik*, 75 (1932).

[104] J. M. Cook, *The Mathematics of Second Quantization* (Thesis, The University of Chicago, 1953).

the known reality of physical facts conditional on a human observer. This holds of the numerous subjectivist distortions of the
Einsteinian relativity. (2) The interpretation, even if free of subjectivist tendencies, must not be physically crippling, i.e., disregard the
physical reality of a scientifically explored aspect of the universe. In
this sense, an interpretation of quantum mechanics which construes
a measurement as a physical interaction between object and instru-
ment but makes the reality of quantum physical events (i.e., of facts
consisting in the circumstance that a physical quantity has a particular value for a particular physical system) conditional on an
actual measurement can be shown to be physically crippling. This
seems to hold of Blokhintsev's interpretation of quantum mechanics. (3) The interpretation must guarantee the empirical testability
or verifiability in principle of the laws which constitute the reinterpreted theory, according to our analysis of the concept of law of
nature. The interpretation of the quantum physical relativity of
time on Einsteinian lines would be physically crippling. The result
of my aforementioned study shows clearly that only an interpretation formulated in an occupation-number type of field formalism
would meet the essential, epistemological requirement of empirical
verifiability in principle. Accordingly, a reinterpretation of the relativity of time in quantum electrodynamics is indispensible. On
closer examination, quantum electrodynamical arguments show
both the necessity of a new interpretation of time relativity and
suggest how it can be established.

Let us first outline a consideration which applies to both classical and quantum physical relativity. The conventional terminology
which I have been using in this paper puts on a par the relativity of
time and space. This is extremely misleading since it makes little
sense to state the dependence of relativistic time on frames of
reference. Nor would it be physically meaningful to attribute any
other property to time. There is no invariantly meaningful split of
the world continuum into time and space components. Statements
about time and space components of the world have therefore no
meaning if they are interpreted literally. The proper time-interval
separating two world points is, however, intrinsically meaningful
and so is the spacelike interval. The world continuum should
therefore be axiomatized in terms of Minkowski-intervals. In this
respect, there is no symmetry between the prerelativistic concepts of

space and time, and the temporal and spatial components of Minkowski's world. I therefore suggest that the world continuum should
be construed as a four-dimensional manifold metricized in terms of
Minkowski's intervals. Both prerelativistic concepts of space and
time would have to be abandoned. The relativity of any physical
quantity, including timelike and spacelike intervals, would then
consist in the covariance of the projection of this quantity onto any
associated pair of a three-dimensional spacelike and one-dimensional timelike components of the world continuum.

An axiomatization of the world continuum metricized by Minkowski intervals will be meaningful both within and without quantum physics if it contains a single physical relation undefined within the axiomatic system but provided with a physical meaning by
appropriate extrasystemic criteria, e.g., of the type Bridgman called
"operational definitions." In the axiomatic system outlined in the
two papers[105] I have published in 1965 and 1966, the crucial
physical concept which discharged this function was that of "collision connectibility." In view of the above remarks, it seems to me
that resorting instead to the relation of "indeterministic causality"[106] would be definitely preferable. The main reasons for so
reaxiomatizing quantum relativistic time are the validity of the
"Principle of Indeterministic Causality" in relativistic quantum
field theories and the circumstance that the new axiomatic system
would not be open to the objection of assigning a strategic role to a
corpuscular concept. I hope to present shortly a full account of this
axiomatic system and to show that it meets the three aforementioned conditions.

In this section, my discussion of temporal relativity dealt almost
exclusively with the ontological aspect of the latter. The ontological
version of the relativity of time comes to the statement that the four-
dimensional world continuum can be split in infinitely many equivalent ways into a temporal and spatial component. The temporal
nature of any of these components is obviously relative to a particular split of the continuum. The epistemological version of temporal
relativity deals with the invariance of all laws of nature under a
transformation pertaining to the Lorentz group. This epistemolog-

---

105 H. Mehlberg, cf. Ref. 94 and 100.

106 H. Mehlberg, cf. Ref. 101 and 102.

ical version has been profoundly modified in quantum theory. Thus, in prequantum relativity, the transformation of spatio-temporal coordinates under the Lorentz group sufficed to guarantee the invariance of the relevant laws of nature. In quantum theory this is no longer the case, in view of a result of I. M. Gel'fand to which I have just referred. Because of shortage of space, it is impossible to discuss here the epistemological aspects of temporal relativity which transcend the ontological issue of relativity.

VI. *The Reality of Time in Quantum Physics*

During the first decades of the development of quantum physics it was often stated that the concepts of time and space are intrinsically inapplicable at the quantum level, even when no doubt was implied as to the validity of these concepts in the domain of classical physics, both relativistic and prerelativistic. In the opening considerations of this study, I have pointed out that, at present, there are compelling reasons for identifying the set of physical objects governed by quantum laws with the set of all physical objects. Consequently, the unreality of quantum physical time would be equivalent to the universal unreality of physical time. This pervasive unreality is certainly incompatible with the sum total of observational, scientific findings presently available. However, in view of the philosophical and scientific significance of the problem of time reality, I shall briefly outline the specific relevance of quantum physics to the ontological version of this issue.

To avoid controversial assumptions, I shall not consider as synonymous the terms "physical reality" and "existence." The mathematically demonstrable existence of any object, e.g., of the infinite set of prime numbers, familiar and pleasant to Greek mathematicians, is as indubitable as the irrationality of the square root of 2, both familiar and shocking to these mathematical ancestors of whom we are justifiably proud. I shall assume that the physical reality of an object X follows from the possibility to obtain observational results such that the existence of X is a logical consequence of these results on the condition that the existence of X does not follow from any set of mathematical axioms[107] (e.g., the axioms of

---

[107] H. Mehlberg, "The Present Situation in the Philosophy of Mathematics," *Synthese*, 14 (1962).

PHILOSOPHICAL ASPECTS OF PHYSICAL TIME

Brouwer and Heyting, or of Whitehead and Russell, or of Hilbert and Bernays). The converse of this statement is far from certain and will not be used in the discussion of the physical reality of quantum theoretical time.

No answer to the question "Is time physically real"? follows from the above three classical sets of axioms for the whole of mathematics. Hence, an affirmative answer to this question would be justified if this answer were shown to follow logically, however indirectly, from reliable, observational findings. In the sequel I shall briefly indicate the logically compelling reasons for the deducibility of the existence of time from crucial, quantum physical theories. More specifically, the following quantum theories were investigated in this context: (1) The nonrelativistic theory of quantum mechanics which still plays a vital role in physical science, in spite of its obvious and crippling limitations. (2) The relativistic theory of quantum electrodynamics, to be used once more as illustrating relativistic quantum field-theories[108] and relativistic quantum theories of elementary particles.[109] The fact that quantum electrodynamics accounts only for some types of elementary particles and is intrinsically incapable of accounting for most of these particles, does not affect the conclusions concerning the physical reality of time which we shall reach by examining the logical structure of quantum electrodynamics.

I shall start my discussion by pointing out that the argument most frequently used in support of the unreality of nonrelativistic, quantum mechanical time was eventually shown to be fallacious, and remains fallacious in relativistic quantum electrodynamics. I have in mind Heisenberg's Indeterminacy Principle, usually construed as implying the nonexistence of a specific value of a dynamical variable for a physical system whose quantum-state is not an eigen-state of the hypermaximal Hermitean operator representing this variable in the formalism of the theory. It is obvious that the time variable is not represented in quantum mechanics by an operator. Time, in physical parlance, is a "c-number," not a "q-number." The validity of the Indeterminacy Principle for the pair

---

[108] P. A. M. Dirac, *Lectures on Quantum Field-Theory* (New York: Academic Press, 1966).

[109] J. Bernstein, cf. Ref. 9.

of dynamical variables consisting of energy and time is therefore not derivable from quantum mechanical axioms although attempts were often made to substantiate the relevance of Heisenberg's Principle to this case by referring to allegedly cogent, "ideal experiment" types of arguments. The cogency of these arguments is only apparent, however. All that can be proved is a remote analogy to Heisenberg's Principle, explored by L.T. Mandel'shtam and I. Ye. Tamm.[110] The determinacy of the energy of a system at a specific time moment is demonstrably compatible with quantum mechanics.

In quantum electrodynamics, analogues of the Indeterminacy Principle, i.e., the non commutativity of field-operators representing measurable quantities, have been established in several, significant cases including processes which involve elementary particles, e.g. pair creation.[111,112] However, as convincingly presented in the treatise of A. I. Akhiezer and V. B. Berestetskii, the noncommuting-operators involved in these indeterminacy-relations have nothing to do with time. Once more, in quantum electrodynamics, time is represented by a "c-number," and its reality status is exactly similar to the status of time in prequantal theories.

In addition to the irrelevancy of the Uncertainty Principle to the reality of nonrelativistic and relativistic time in a universe governed by quantum laws, it should be pointed out that the existence of time is demonstrably deducible from both quantum mechanics and quantum electrodynamics. H. Poincaré used such an existence argument in connection with the physical reality[113] of atoms and molecules which then seemed as problematic to scientists interested in the foundations of physics as the physical reality of time in quantum theory probably seems to today's investigators with similar, theoretical interests. Poincaré answered the question of whether atoms and molecules really exist by pointing out that since they can be counted they obviously could not be nonexistent.

110 L. I. Mandel'shtam and I. Ye Tamm. *Izviesta Akademii Nauk S. S. S. R.* 9, 48 (1945). (Bulletin of the Academy of the U.S.S.R.).

111 A. I. Akhiezer and V. B. Berestetskii, cf. Ref. 28.

112 S. S. Schweber, *An Introduction to Relativistic Quantum Field Theory* (New York: Harper & Row, 1961).

113 H. Mehlberg, "The Problem of Physical Reality," in *Quantum Theory and Reality*, Ch. 2, (New York: J. Springer-Verlag, 1967).

The countability argument for physical existence was then based upon the theory of Brownian Movements. An exactly similar and equally cogent argument supporting the existence of time can be derived from quantum electrodynamical (and quantum mechanical) laws concerning transition probabilities.

It is advantageous to use, at this juncture, a most remarkable alternative formulation of quantum electrodynamics due to Feynman.[114] Instead of the aforementioned set of fundamental quantum electrodynamical equations, the Feynman formulation of this relativistic quantum theory consists of a single equation which determines the probability of the transition of the compound quantum system consisting of an electromagnetic field and the quantum particles present in this field during a definite interval of proper time. The transition involves an initial state characterizing the compound system at the beginning of the proper time interval and any particular state which would be ascribable to the compound system at the final instant of the interval. The validity of relativistic time is apparent from the very formulation of this electrodynamical axiom. Moreover, the probability $\underline{p}$ that during this interval a transition from the initial to the final state will occur is physically interpreted in terms of "relative frequencies" as follows: To begin with, we consider a great number N of nonoverlapping proper time intervals of the same magnitude and the same initial state of the compound system. Then, the number of those proper time intervals during which the transition to the same final state occurred is equal, by definition, to $\underline{p}$.N. To use, once more, Poincaré's way of speaking: quantum electrodynamics provides for the countability of those proper time intervals during which a specifiable change of a compound quantum system took place. Nonexisting proper time intervals could not possibly be counted.

The formulation of electrodynamics just referred to is an important example of Feynman's so-called "spatio-temporal" approach to quantum theory. The Feynman approach[115] is actually a third legitimate "picture" of quantum theory, distinct from the Schrödinger and the Heisenberg pictures and, philosophically, appreciably

114 R. P. Feynman, cf. Ref. 15.
115 *Ibid.*

less opaque. Its relevance to the problems of time reality and relativity is discussed in a separate work of mine.[116]

Similarly to the symmetry and the relativity of quantum physical time, its reality is established by presently available quantum laws, rather than by individual observations of individual facts. However, only intrasystemic laws of logical rank 1 were relevant to the first two problems. On closer analysis, we would find out that physical laws of higher logical rank are relevant to the reality of time. A law of rank 2 states that no law of rank 1 which governs quantum electrodynamical processes and is expressed in an inertial coordinate system would cease to be valid if reformulated in another inertial coordinate system. The reality of proper time follows from this law of rank 2, for if there were no four-dimensional continuum then no law dealing with it could possibly be true.

HENRYK MEHLBERG

THE UNIVERSITY OF CHICAGO

---

[116] H. Mehlberg, *Philosophy and Atomic Theory*, forthcoming, University of Chicago Press.

CAUSAL THEORIES OF TIME

There is a family of theories of time which attempt to construct the structure (or at least the topology) of space-time on the basis of the notion of causality, or of some notion closely related to that of causality. In the last fifty years or so such attempts have been inspired by the special theory of relativity. Consider a point A in space-time. All light signals arriving at or departing from A lie along generators of the four-dimensional analogue of a double cone, which may be called the "light cone" of A. Then it is a consequence of the special theory of relativity that any event which is causally connected with A must lie within the double light cone. No event outside the light cone is causally connected with A. This suggests the possibility that there is some deep connection between causality and the structure of space-time, and even that the former may be used to explain some or all features of the latter.

In the last paragraph I have been using the words 'causality' and 'causally connected' in a fairly loose manner. The proposition that in special relativity causal influences lie only within light cones is almost trivial, in the sense that it is assumed that any causal influence is explained physically by a physics which conforms to the kinematics of special relativity. The proposition is saved from triviality by the tacit assumption that physics which conforms to relativity is the only physics that there is (or ought to be). This assumption has plenty of empirical evidence to support it.

If we know that an event C is the cause of an event E, then we know that there exists at least one light cone, such that C lies in its earlier half and E lies in its later half. It thus looks easy to define a relation of earlier and later in terms of that of cause and effect. This was the basis of Reichenbach's construction in his book *The Philosophy of Space and Time*.[1] Reichenbach's construction runs

---

[1] Hans Reichenbach, *The Philosophy of Space and Time* (New York: Dover, 1958), originally published in German in 1928 under the title *Philosophie der Raum-Zeit-Lehre*.

up against certain difficulties, which have been pointed out by
Henryk Mehlberg in his "Essai sur la Théorie Causale du Temps"[2]
and by Adolf Grünbaum.[3] In the absence of a prior notion of
earlier and later how can we distinguish cause from effect, especially
in view of the time-symmetry of the laws of nature? (If we may neglect the recent discovery of a rather recondite violation of time-symmetry in physics, the 2π decay of the $K_2{}^\circ$ meson.) Reichenbach attempted to answer the objection by means of his so-called mark
method, which is essentially an appeal to irreversible causal processes. This is objectionable, because it is desirable to separate the
discussion of the topology of space-time from the question of the
directionality of physical processes, since (as Reichenbach himself
has argued) this directionality apparently depends on statistical
considerations. I shall not reiterate the detailed objections to Reichenbach's mark method: these have been well made by Mehlberg and
lucidly recapitulated by Grünbaum.

Grünbaum's version of the causal theory of time (like Mehlberg's) makes use of a symmetrical concept of causal connectedness
("k-connectedness"), instead of an asymmetrical concept of cause
and effect. His construction therefore avoids the circularity which
might arise from the fact that the concepts of cause and effect
already presuppose those of earlier and later.

How do we deal with points of space-time which are not occupied
by causes or effects? One way which probably will not do is to use
for our construction not the concept of causal connectedness but the
concept of causal connectibility. This is because of the modal nature of the concept of connectibility: the notion obviously depends
on that of physical possibility, which in turn refers back to the laws
of nature. If these laws of nature themselves presuppose the very
structure of space-time which we are seeking to elucidate by means
of the notion of causal connectibility we are clearly involved in a
vicious circularity. Grünbaum therefore seems to be quite right in
basing his construction on the notion of connectedness, not that of
connectibility. The question about points of space-time which are
not occupied by causes or effects can perhaps be answered from a
deterministic point of view by saying that we need not consider

---

[2] *Studia Philosophica* I (1935) , pp. 119-258, and II (1937) pp. 111-231.

[3] See Adolf Grünbaum, *Philosophical Problems of Space and Time* (New York:
Knopf, 1963) , Chapter VII.

such points: on a relational theory of space-time we need analyse
only spatio-temporal statements which are about actual events. It is
difficult, however, to see how to deal with points of space-time
which are occupied by events which are neither effects nor causes of
other events. Surely, even on a relational theory of space-time, such
an event has to be located somehow (supposing that there are such
events). Alternatively, we might, as Mehlberg does,[4] take it that
there will be events everywhere, for example the event of the
gravitational field (or the electrical field) taking a certain value.

Let us concede, then, that causal connectedness, though not causal connectibility, will do the trick. Nevertheless my chief complaint
against the causal theory of time still stands, because even the
notion of causal connectedness would appear to be a disguisedly
modal one, or at least an intensional one. Does it not, on the face of
it, seem to be very odd to want to reduce a geometrical concept
(and hence presumably one whose axiomatisation can be carried
out in an extensional language) to a philosophically questionable
(and probably at bottom intensional or modal) notion like that of
causality? Surely space-time need be no less transparent to the
intellect than is geometry. Is not geometry a respectable mathematical business, with no need of dubious metaphysical notions like
that of causality? To this a proponent of the causal theory of time
may reply that my objection neglects the difference between pure
geometry and applied geometry, and it is the purpose of the causal
theory of time to provide correspondence rules which tie down pure
geometry (or perhaps at least its nonmetrical part) to reality, thus
converting it to a physical geometry.

An additional consideration might be put forward by a proponent of the causal theory of time. He might wonder why I should be
squeamish about using causal connectedness as a primitive in a
physical geometry. It might be said that the notion of causal connectedness is no more objectionably modal than is the operator 'it is
a law of nature that . . .', and that it is necessary to retain this
operator in the special theory of relativity, whose chief principle is
the assertion that the laws of nature retain the same form under
transformations from one set of inertial axes to another. However, I
prefer to take the view that the principle of the invariance of the

---

4 *Op. cit.* II, pp. 163-64.

laws of nature under Lorentz transformations is a heuristic maxim
for construction of the theory rather than an axiom of the theory
itself. Alternatively it can be interpreted as a metatheoretical
statement (a statement *about* the theory, not *of* the theory). Thus
Maxwell's equations can easily be seen to be Lorentz invariant, and
Lorentz invariance guided Einstein's search for the relativistic laws
of mechanics which replace Newton's laws. Maxwell's equations and
the relativistic laws of mechanics do not themselves contain the
expression 'it is a law of nature that . . .' or any equivalent
expression. It would indeed be odd if a physical theory could not be
stated without the help of the apparently *metalinguistic* concept of
'law of nature'.

Similarly, one should hope that any physical theory can be
formalised without strong conditionals or the notion of natural
necessity. On the view which I should support, statements of natural necessity are disguised metalinguistic statements to the effect
that certain truth-functional conditionals follow from fairly fundamental laws of nature. Trivially, then, we may say that any *fundamental* law of nature expresses a natural necessity because the law is
of course deducible from itself. Even "merely empirical" generalisa-
tions may sometimes be said to express a natural necessity in the
sense that we hope or believe that they may one day be shown to
follow from some suitable theory. Thus physical necessity is a metatheoretical notion, not a theoretical notion of physics.

As against this view of natural necessity it is sometimes contended that strong conditionals are essential to science, and are not
disguised metalinguistic statements. Thus Newton's first law of
motion (that if any body is not acted on by a force it moves with
uniform velocity in a straight line) is probably no more than
trivially true if it is interpreted in the sense of the weak conditional,
because probably there are no bodies in the universe which are
acted on by no force at all. The reply to this is that Newton's first
law of motion is a dispensable luxury, because it is a special case of
the second law, which is true in a nontrivial way. Moreover even if
a law should be trivially true it is then at least *true*, and so it can do
no harm to the theory. (No falsehood can be deduced from a truth.)
It is true that Newton's first law does seem to occur in a nontrivial
way in some presentations of special relativity. In Newtonian mechanics we can define an inertial system as one according to which

the components of momenta of mutually attracting bodies balance out, but in special relativity this is not possible, because of the arbitrariness of simultaneity of spatially distant events. It might therefore be thought necessary to define an inertial system by reference to Newton's first law, expressed as a strong conditional. However in principle this course can be avoided by defining an inertial system by reference to collision phenomena. In practice, of course, this method is too unwieldy: it would be too bad for humanity if in celestial mechanics collisions between the heavenly bodies occurred often enough for this purpose. And so in practice we will have to specify an inertial system by taking a Newtonian one as a first approximation. Nevertheless this method does not require an appeal to natural necessity or the use of strong conditionals.

My position, then, is that a physical theory should be based on a purely extensional language, and the predicates '. . . is necessary' and '. . . is necessary for . . .' should not occur in it. (For detailed support of this philosophical stance, let me refer to the work of Quine, especially his *Word and Object*.[5]) Now if we look at Grünbaum's exposition of the causal theory of time, we find that he does make use of the concept of necessity. This is notwithstanding the fact that his theory is based on the notion of causal connectedness and not on the obviously modal notion of connectibility, which he rightly rejects. He recognizes, moreover, that causal connectedness can not be defined in terms of necessary and sufficient conditions for the occurrence of events, for reasons similar to those which are familiar from Nelson Goodman's work on subjunctive conditionals.[6] (I myself would be happy to elucidate causal connectedness metalinguistically, as suggested earlier in this article, but clearly what Grünbaum needs is an intratheoretic, not a metatheoretic, concept of causal connectedness.) Grünbaum therefore introduces the concept of causal connectedness as a *primitive*. So far (perhaps) so good. But he does need the concepts of 'necessary' and 'sufficient' in his definition of 'n-quadruplet'. This last notion is introduced for the purpose of defining the notions of betweenness and

5 W. V. Quine, *Word and Object* (New York: John Wiley, 1960).

6 See Grünbaum's remarks on p. 609 of his paper "Carnap on the Foundations of Geometry" in P. A. Schilpp (ed.), *The Philosophy of Rudolf Carnap* (Open Court: La Salle, Illinois, 1963), pp. 545-684, and his reference to Goodman's *Fact, Fiction and Forecast* (Cambridge, Massachusetts: Harvard University Press, 1955).

separation closure. Thus the quadruplet of events ELE'M is said by Grünbaum to be an n-quadruplet if and only if, "given the actual occurrence of E and E', it is necessary that either L or M occur in order that E and E' be k-connected, L and M being genidentical with E and E', . . ."[7]

Mehlberg[8] defines the notion of one event "acting on" another in a more apparently extensional fashion. According to him, an event A acts immediately on an event B if for every event A' which is intrinsically similar to A there corresponds an event B' which has the same intrinsic properties as B and the event B' coincides partially with A'. Here the intensionality resides in the notion of one thing's being intrinsically similar to another and in the notion of an intrinsic property: this can be checked by consulting Mehlberg's definitions of these notions. It is also not clear to me that Mehlberg's definition of 'immediately acts on' avoids Goodman-like objections. However I have to confess that my understanding of Mehlberg's complex theory of decompositions of events is as yet defective. It does seem to me, however, that his concept of an event is an implicitly intensional one. Let me now, however, pass on to discuss another type of objection to certain causal theories of time.

Following Reichenbach, who derived it from K. Lewin,[9] Grünbaum speaks of causal chains of events as "genidentical." He goes on to "utilise the property of causal continuity possessed by genidentical causal chains"[10] and asserts that associated with any two genidentically related events there are certain classes of events which are genidentical with these first events, and that these classes have the cardinality of the continuum. It is not necessary here to go into the full details of Grünbaum's assertion, which is a little bit complicated because of his laudable desire to avoid prejudging the question of whether time is topologically open (like a Euclidean straight line) or closed (like a great circle on a sphere). However one of the classes of events in question is the class of events which, in the case of the topological openness of time, could be said to be *between* the two originally given events.

---

7 *Philosophical Problems of Space and Time*, p. 194.

8 *Op. cit.*, II, p. 153.

9 See Reichenbach, *op. cit.*, p. 142, footnote.

10 Grünbaum, *Philosophical Problems of Space and Time*, p. 194.

Now it seems to me to be a defect in a theory of time that it should have to depend on the assumption of continuous classes of events. Certainly according to classical physics there will be a continuous set of genidentical events between any two genidentical events, but it is hard to make much sense of this in the context of quantum mechanics (unless a value of the ψ-function perhaps counts as an event) .[11] In saying this I do *not* want to fall into the error, so well castigated by Grünbaum,[12] of supposing that quantum mechanics allows of a discrete space-time. Quantum mechanics uses continuous mathematics and a continuous geometry, like any classical theory. Indeed it is not at all clear what a physical geometry based on a discrete space-time would look like, especially in view of the well known incommensurability of certain geometrical ratios and in view of Zeno's paradox of the stadium. Fully allowing Grünbaum's point here, I wish only to query how the postulation of a continuous set of genidentical events would work out in the context of quantum theory. An obvious difficulty comes from the breakdown of causal connectedness in quantum mechanics. The difficulty shows itself in the absence of definite trajectories, which is evidence of an uncertainty as to just what is supposed to be the set of events which would make up a genidentical causal chain. Perhaps Grünbaum is offering the causal theory of time as a construction which works within the context of classical physics (as modified by special relativity) . However we may well doubt the utility of a construction which works only for classical physics.

The general theory of relativity also would seem to make big trouble for the causal theory of time. This is because the causal theory of time is a form of the relational theory of time: it defines betweenness relations (or separation closure relations) only for quadruplets of actual events. However the question of whether or not a relational theory of time will work for general relativity boils down to the question of whether relativistic cosmologies can accommodate Mach's principle, and this is a rather open question.

---

11 A difficulty here is that the Ψ-function relates to points of a many-dimensioned Hilbert space, whose coordinates are given by complex numbers, and not to our ordinary space-time.

12 See Adolf Grünbaum, *Modern Science and Zeno's Paradoxes* (Middletown: Wesleyan University Press, 1967) , pp. 109-14.

(Grünbaum has indeed himself argued to this effect.[13]) The above objection is probably not valid against Mehlberg's version of the causal theory of time, because for him there are events everywhere, though it is hard to see how he could accept the complete geometrisation of the gravitational field and the theoretical possibility (in some cosmologies) of a universe devoid of matter and yet with a certain space-time structure. In fact Mehlberg has conceded this sort of point in a later paper, in which he rejects the causal theory of time.[14]

As we remarked earlier, modern attempts at a causal theory of time derive their inspiration from special relativity, which rejects action at a distance and implies that all causal influences (all world lines, all signals) lie within light cones. This implication is, however, extratheoretical rather than intratheoretical. If it is not a heuristic maxim it is a theorem in the metalanguage of physics. It is instructive to consider two space-time objects, $B'AB$ and $C'AC$, as in Figure 1, such that the former lies entirely within the light cone whose vertex is A and the latter lies (except for the point A) entirely outside the light cone. Clearly we should allow $B'AB$ to determine a world line, whereas we must not regard $C'AC$ as lying along a world line but must interpret it as a brief but widely occurring disturbance. The structure of classical physics is such that a sufficiently detailed knowledge of a sufficiently large "time slice" through B' would enable us to deduce the event B, whereas no amount of knowledge of however large a "space slice" (through C') would enable the event C to be deduced. This is what the causal grain of the world, as determined by world lines, comes to. These are the insights which give rise to the causal theory of time, and they can be preserved, in the present *metatheoretical* form, by a physics which postulates space-time independently of the notion of causal connectedness.

---

[13] Adolf Grünbaum, *Philosophical Problems of Space and Time*, Chapter XIV.

[14] See Henryk Mehlberg, "Relativity and the Atom," in Paul Feyerabend and Grover Maxwell (eds.), *Mind, Matter and Method, Essays in Philosophy and Science in Honor of Herbert Feigl* (Minneapolis: University of Minnesota Press, 1966). See especially pp. 484-86.

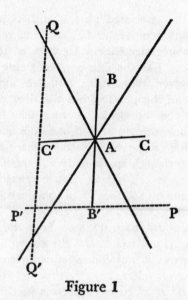

Figure 1

The insights which give rise to a causal theory of time can then be preserved by a philosophy of science which simply postulates a structure of space-time, and then together with the associated physics and its *metatheory* deduces causal relations between events in the world. In postulating that events are related as in accordance with such and such a geometry we can even keep the options open as between a relational and an absolute interpretation of this geometry. In the case of an absolute theory of space-time the spatio-temporal relations will directly relate not events but space-time points, which will have the status of theoretical entities of physics. In the case of a relational theory the spatio-temporal relations will relate events in the world directly. The primitives of such a geometry might perhaps be those of conical order, as in the theory of A. A. Robb[15], and would not involve the notion of causal connectedness.

In this way, instead of "constructing" the (nonmetrical) structure of space-time from a relation of causal connectedness we deduce relations of causal connectedness and nonconnectedness from the

---

[15] A. A. Robb, *A Theory of Time and Space* (Cambridge: Cambridge University Press, 1914).

theory (including the associated physics and its metatheory). We can also deduce spatio-temporal relations from relations of causal connectedness or nonconnectedness. In view of the preceding remarks this might look like the fallacy of affirming the consequent. This is not so, however. We deduce that if E and E' are causally connected then one of them lies within the light cone of the other by transposition of the proposition (which may be deduced from the theory plus its metatheory) that if one of E and E' does *not* lie within the light cone of the other then they are *not* causally connected. In this way, though spatio-temporal relations are logically prior to causal ones, we can indeed deduce certain spatio-temporal relations from certain causal ones. This, once more, is the grain of truth in a causal theory of time.

To conclude, then, my main qualms about the causal theory of time are as follows: (1) To elucidate the concept of space-time in terms of the concept of causal connectedness seems to be to elucidate the comparatively clear by reference to the comparatively unclear. (2) Even if causal connectedness is itself taken as a primitive, causal theories of time seem to need modal or intensional concepts somewhere or other. (3) It is difficult to see how the causal theory of time is applicable to theories which allow for the existence of events which are neither causes nor effects of other events. It at least *seems* to me that I can consistently envisage a universe of purely random events spread out through space-time. The objection that only if such events were causes or effects of other events could they in fact be located ought, I think, to be dismissed as too verificationist. On p. 88 of his "Reply to Hilary Putnam"[16] Grünbaum remarks that "spatially separated events can sustain physical relations of one kind or another only because of the presence or the absence of their actual or at least physically possible connectibility." It seems to me that if 'physical relations' here means 'causal relations' Grünbaum's assertion is doubtful, or at least dependent on certain philosophical assumptions. It can not be used as a premiss with which to refute someone who does not accept a causal theory, because such a person could say that some physical relations (namely spatio-temporal ones) could exist even in the

---

16 In R. S. Cohen and M. W. Wartofsky (eds.), *Boston Studies in the Philosophy of Science*, Vol. V (Dordrecht, Holland: D. Reidel Publishing Company, 1968).

absence of causal connectibility. Finally, I do not think that I have conclusively *disproved* any causal theories of time. I would claim only to have expressed certain qualms about causal theories of time and doubts about the motivation for such theories. For example I have by no means adequately discussed Mehlberg's stimulating and detailed treatment of the subject. It is quite possible that the qualms I feel about causal theories of time are due more to defects in my own thinking than to defects in the theories. Even so, since I suspect that others may be liable to feel similar qualms, I hope that it has been of some value to have ventilated the matter, however unsatisfactorily.[17]

J. J. C. SMART

UNIVERSITY OF ADELAIDE

---

[17] See also a paper by Hugh M. Lacey, "The Causal Theory of Time, A Critique of Grünbaum's Version," *Philosophy of Science* 35 (1968), pp. 332-54, which appeared since the present one was submitted for publication.

'HERE' AND 'NOW'

In my book, *The Language of Time*[1], it was argued that the distinctions of past, present and future are objective. The over-all structure of the argument was as follows: (i) 'now', as well as other temporal demonstratives, although not designating a sensible property, has an informative role in our language, and for this reason temporal demonstratives cannot be eliminated without loss of information; (ii) 'now' is "semantically" objective but "pragmatically" subjective, i.e. a sentence containing a word such as 'now' is not used to refer to any mental event or linguistic tokening of itself (as contended by Russell and Smart respectively), but the rules controlling the use of such a sentence for making a true statement do specify that a specific kind of temporal relation must obtain between the tokening event and the event reported by the statement made through this tokening event; (iii) the use of 'now' is not selective, picking out some moment of time or class of simultaneous events from amongst others that might have been designated in its place; and, (iv) the realistic commitments of ordinary language or common-sense concerning 'now' are not, and could not be, inconsistent with any empirical fact or physical theory. Points (i) - (iv), taken in conjunction, go a long way toward establishing that there is nothing mind-dependent or linguocentric about *now*, that events could be past, present and future and change in respect to these distinctions even in a world devoid of perceivers or language-users.

My thesis concerning the objectivity of *the present* faces a very disturbing *ad hominem* rebuttal. Those who have held *the present* or *now* to be objective, such as C. D. Broad and myself, have not been willing to countenance an equally objective *here* in nature, no less a *here* which "shifts" from place to place. Unfortunately, 'here' also seems to satisfy (i) - (iv) above.[2] If these establish the objectivity

---

[1] Richard M. Gale, *The Language of Time* (London: Routledge & Kegan Paul, 1968). Hereinafter cited as LT.

[2] It might be thought that 'here' fails to satisfy (iii) because, when used in conjunction with an act of ostension, the speaker can select which place is

of *now*, they do likewise for *here*—a most disturbing consequence. To avoid this consequence it will be shown that there are crucial respects in which our use of 'here' but not 'now 'is selective, and as a result 'here' does not fully satisfy condition (iii). Before doing this, it will be shown that 'now' and 'here' are disanalogous in other important respects. This will be done by showing that there are important conceptual truths concerning one of these terms that have no analog concerning the other. A given statement will be said to have no *true* spatial (temporal) analog if its analog is either nonsensical or has a different logical status, e.g. is contingent while the original statement is necessarily true.

First, a few preliminary remarks about what will count as a spatial analog to a temporal statement, and vice versa. In forming such an analog we must substitute for every spatial term a corresponding temporal term, and vice versa. As a result, the analog of the analog of a given statement is this statement. This principle of double analog parallels the familiar principle of double negation. Before attempting to form the analog to a given statement it is important that all of the spatial and temporal commitments of this statement be made explicit in such a form as to permit us to interchange spatial and temporal terms. An example of a statement that runs afoul of this requirement is:

(T_1) A future utterance of 'now' will denote a different time. To which the spatial analog supposedly is:

(S_1) A future utterance of 'here' will denote a different place. It is clear that (S_1) is not a *true* spatial analog to (T_1), the reason being that (T_1) is necessarily true while (S_1) is contingent. That (T_1) has no spatial analog is supposed to show that there is no spatial analog to temporal becoming—the "shift" of *the present*.

---

designated. In this use, 'here' functions like the demonstrative 'this'. There is no analogous use of 'now', since different times, unlike different places, do not coexist, and as a result only in the latter case is there a field of objects presented to us out of which we can make our selection through ostension. But a *naked* use of 'here'—one not accompanied by an act of ostension—is no more selective than is the use of 'now'; for just as 'now' cannot help but indicate a time which at least includes the time at which it is tokened so 'here' cannot help but indicate a place which at least includes the place occupied by the speaker. This paper will be concerned exclusively with the naked use of 'here'. If *here* has a different ontological status from *now* this should reveal itself in a significant array of disanalogies between "now" and "here," even when only naked uses of 'here' are considered.

The difficulty with (T_1) is that it fails to make explicit all of its temporal commitments, as does this more perspicuous version of it:

$(T_1)'$ An utterance of 'now' later than now denotes a different time.

But there is a genuine spatial analog to this, viz.:

$(S_1)'$ An utterance of 'here' removed from here denotes a different place.

Both statements express conceptual truths. With these pitfalls in mind we can now get down to business.

The significant disanalogies between 'here' and 'now' do not lie on the surface of our language. They can be flushed out only by noting a difference in the way in which these terms are involved in some of the basic concepts by means of which we think and talk about the world. Two such basic concepts will be considered—*objects* and *agency*—The latter alone bringing out the sense in which 'here' but not 'now' is selective and therefore not objective.

1. Disanalogies Between 'Here' and 'Now' Concerning Objects

An obvious difference between 'here' and 'now' is connected with the conceptual truth that an object cannot at the same time be both here and there but can at the same place be both now and then. This surface disanalogy is rebutted by the claim that at a given time an object can be both here and there, provided it fills here and there as well as the space between, just as analogously it can at a given place exist both now and then provided it exists both now and then as well as at the time between.[3] The next move in this familiar dialectic is to point out that it is only part of the object that exists here and there at the same time, whereas it is not only part of the object that exists both now and then at the same place. In short, objects appear to have spatial but not temporal parts.

With this in mind, we can try to reformulate these conceptual truths:

(S_2) An object cannot wholly occupy both here and there during the same interval of time.

It is important to make clear what is meant by an object wholly

---

[3] For a defense of this see Richard Taylor, "Spatial and Temporal Analogies and the Concept of Identity," *Journal of Philosophy*, 52 (1955); and "Moving About in Time," *The Philosophical Quarterly*, 9 (1959).

occupying a given space during a certain interval of time. The following is a definition of what is to be meant by this:

(D$_s$) An object O wholly occupies a region of space S during an interval of time T if, and only if, every spatial part of O existing during T is contained in S and there is no sub-region of S not filled during T by some spatial part of O.

A more simple way of formulating the *definiens* of (D$_s$) is that during T the spatial boundaries of O and S coincide. Given this sense of 'wholly occupy' there can be no doubt that (S$_2$) is a conceptual truth.

The temporal analog to (S$_2$) is:

(T$_2$) An object cannot wholly occupy both now and then at the same region of space.

Before we inquire as to whether this is a conceptual truth, we must ask what could be meant by an object "wholly occupying a time at a space"—a most bizarre notion. Let us try to find some meaning for it by forming the temporal analog to definition (D$_s$).

(D$_t$) An object O wholly occupies an interval of time T at place S if, and only if, every temporal part of O occurring at S is contained in T and there is no sub-interval of T not filled at S by some temporal part of O.

This is not much help, since it employs the equally strange notion of a "temporal part" of an object.[4] While we do speak of the spatial parts of an object, we do not speak of its temporal parts. E.g. if you had an unobstructed view of some object, such as a woman, during a sub-interval of her total life-span you would not say that you saw only part of her, as you would if you could see only her toes or fingers due to her heavy Moslem veils. It is not enough, however, to point out that we do not ordinarily speak of the temporal parts of an object; some justification or explanation must be given of these linguistic facts. The following will show that there are good reasons why we do not speak about the temporal parts of an object.

The discussion will be restricted to substantial objects having an internal complexity of parts, such that if such an object be divided

---

4 In both of the Taylor articles mentioned in footnote 3, he introduces the notion of the temporal parts of an object by considering an object to be a four-dimensional space-time entity. He fails to notice that this is not the ordinary concept of an object; and, as a result, his arguments to prove that *objects*, supposedly taken in their ordinary sense, can move in time in ways analogous to those in which they can in space, fail through equivocation.

in half neither of these parts would be a substance of the same kind as the undivided object. Such an object will be referred to as a "sortal-object" and will be said to have a "sortal-nature"—the what-it-is. If an object ceases to have its sortal-nature it ceases to be. A sortal-object will be distinguishable, reidentifiable and countable, although the converse does not hold; e.g. a lump of coal satisfies the latter conditions but is not a sortal-object since half of a lump of coal is still a lump of coal. The argument will be to the effect that a sortal-object does not have temporal parts that are analogous to its spatial parts, the reason being that its spatial parts alone are connected with its sortal-nature.

The crucial reason for this disanalogy between space and time is that an object's sortal-nature is dependent, at least in part, upon the way it fills space—its size and shape—but not upon its way of filling time—its history. There is no interesting temporal analog to an object's spatial shape and size. First, consider shape. To be a certain sort of sortal-object, e.g. a man, requires that an object have, within certain ill-defined limits, a certain spatial shape at any given age. It is necessary to relativise an object's shape to its age because some objects, e.g. organic ones, change their shape with their age. The spatial shape is at least a necessary condition for its having a certain sortal-nature, although it is often not sufficient, since the stuff an object is made of might also be relevant. The temporal analog to an object's shape—the area of space it occupies at a given time—would be the times (s) during which it occupies a given place. But an object's time (s) at a given place is unrelated to its sortal-nature.[5] E.g. if it be known that an object occupies a certain building every evening from 8 p.m. to 12 p.m. nothing relevant to the determination of its sortal-nature is known. It could be a man, e.g. the janitor, or the broomstick he always brings with him. If it be claimed that the man and the broomstick do not occupy spaces of the same shape we are distinguishing between their sortal-natures not on the basis of their different "temporal shapes" at a space but on the basis of their different spatial shapes at any given time during their history. We could make up a sense for an object's "temporal shape" by saying that it is the spatial area marked off by a Galtonian photograph of an object's total history, i.e. the spatial

---

[5] I have learned a good deal about this from Judith Jarvis Thomson's excellent article on "Time, Space, and Objects," *Mind*, 74 (1965).

area formed by projecting its spatial coordinates at each instant of its history onto some one spatial cross-section. This strange notion of an object's "temporal shape" still is of no use in determination of its sortal-nature; e.g. a cubical object, such as a die, could have a spherical "temporal shape" in this sense, provided its center-point remained stationary but it rotated around both its vertical and horizontal axes.

Closely connected with the fact that a sortal-object must have a certain kind of shape is the requirement that it possess, at any given age, a certain size, once again within certain ill-defined limits. There are fat men and skinny men, but there could not be any man the size of the solar system. Certainly a sortal-object could not be infinitely or indefinitely extended in one of its spatial-dimensions. How different it is with the way it fills time—its "temporal size." Not only is there no requirement placed upon the length of a sortal-object's duration, other than the trival one that it must endure for a finite time, but there is no absurdity in such an object being infinitely or indefinitely extended into the past or future.[6] It is conceivable that a sortal-object, such as a baseball, would endure indefinitely or infinitely into the future. Those who seek the life eternal do not hunger after the absurd.

There are further disanalogies between the way in which a sortal-object fills space and time.

(T₃) A sortal-object of a certain kind exists during any sub-interval of its history, and a sortal-object of that kind would exist during such a sub-interval even if the sortal-object were not to exist at any time earlier or later than that sub-interval.

Consider a man who exists for a total of 60 years. During any sub-interval of his life, e.g. the second ten minutes of his life, this *man* exists, and his existence during this sub-interval is logically independent of the fact that he exists at times earlier and/or later than this sub-interval. To be a man does not require any particular kind of origin, such as being born of a woman; if so the Frankenstein story would be an absurdity, which most certainly it is not. The spatial analog to (T₃) runs afoul of the counter-factual clause. It would be:

---

6 Unlike a sortal-object, a chunk- or mass-object, such as snow or water, could be indefinitely extended in space. The event or process analog (though they are not perfect) to a chunk-object is a process such as walking or raining. They could go on for an indefinite period of time. Like snow, raining is not internally complex.

(S₈) A sortal-object of a certain kind exists in any sub-region of the space it occupies, and a sortal-object of that kind would exist in such a sub-region even if the sortal-object in question were not to exist at any place surrounding that sub-region.

It is a consequence of the definition of a sortal-object that the counter-factual clause is conceptually false: a sortal-object is such that if it is divided into parts it is not the case that each one of these parts is an object of that kind. A man can be said to exist in the region of space occupied by his arm only because the space adjacent to this region is filled in a rather specific way—by a torso, head, etc. The space occupied by an arm severed from the body is not occupied by a man, not even by the arm of a man.

It might be argued that an object's way of filling time—its history—is directly tied to its sortal-nature, even though there is no interesting temporal analog to the spatial shape and size of a sortal-object. Often the sortal-nature of an object involves that it possess certain dispositional properties; and the only way we can discover what dispositional properties an object has is by observing the way in which it behaves, i.e. its history. This objection runs together the epistemic question of how we *discover* that an object has a certain dispositional property with the ontological question of what it is for an object to have this dispositional property. While we discover that an object has a certain dispositional property by observing how it interacts with other things, that an object has a certain dispositional property does not entail that it actually realizes this property; e.g. it may never be presented with an opportunity for exercising this disposition. There is no contradiction in speaking of a knife which spends its whole history sitting on a mantle as an item of decoration or of a golden ring which never gets dissolved in aqua regia. Even complex biological sortal-objects need not have any specific kind of history in any way analogous to the sort of shape and size they must have. E.g. there is no action (s) or mode (s) of behavior that are constitutive of being a man.

Another objection based on confusing ontology with epistemology goes as follows. Suppose we claim that a particular sapling is an oak, but it subsequently grows palm leaves. This fact about its subsequent *history* shows that it was a palm tree all along. But this shows only that in *finding out* whether the sapling is an oak we take into account facts about its future history. The sapling, however, could have been an oak even if it had no future history, due to its being destroyed. Furthermore, it is because the sapling has a determinate spatial structure that it has the potentiality to grow a certain kind of leaves.

In summary: we are not able to form the temporal analog to conceptual truth (S_2). To do so we must utilize the concept of the temporal parts of an object; however, it was shown that a sortal-object does not have temporal parts analogous to its spatial parts, this resting on the fact that only the latter sort of parts are relevant to its sortal-nature.

A fundamental objection might be raised here. It could be argued that (T_2) is not the temporal analog to (S_2), since not every spatial term in (S_2) was replaced by a corresponding temporal term in (T_2). Specifically, the trouble concerns the use of 'sortal-object' in (T_2) as well as in (S_2). In (T_2) this term should have been replaced by its temporal analog, viz. that of a *sortal-event*. A sortal-event will have an internal complexity of temporal parts that determine its sortal-nature—the kind of event it is—just as analogously a sortal-object is made up of spatial parts that determine its sortal-nature. Examples of sortal-events are running a mile, drawing a circle, and other events which terminate in an "achievement" event.[7] In regard to a sortal-event we can say at exactly what instant a given process qualifies as a sortal-event of a certain kind. Accordingly, (T_2) should be amended to read:

$(T_2)'$ A sortal-event cannot wholly occupy both now and then at the same region of space.

This appears to be just as much a conceptual truth as is (S_2). The first half of a sortal-event, such as a football game, is only part of this event: if you witnessed only the first half of the game it would be correct to say that you witnessed only part of this event, and not because your view was obstructed by a pole or the like.

This objection will be met by showing that a sortal-event is not the temporal analog to a sortal-object. Before doing this, there are two things to be mentioned about this objection. First, it could not be raised consistently by those philosophers who hold that an object *is* an event (a boring one) or group of events, since this objection claims that things (objects) and events are different, although

---

[7] For a fuller discussion, see my article, "Can a Prediction 'Become True'?" in the *Philosophical Studies*, 13 (1962).

analogous, sorts of entities, the former being made up of spatial and the latter of temporal parts. Second, this objection presupposes the truth of my contention that the concepts of *here* and *now* are not involved in an analogous way in our concept of a sortal-object, this being a direct result of the fact that a sortal-object is an essentially spatial concept.

To rebut this objection it will be shown that there are conceptual truths concerning a sortal-object which do not have true temporal analogs concerning a sortal-event. Consider the following pair of statements:

(S₄) A sortal-object can wholly occupy different places at different intervals of time, it being the whole object that exists at each of these two different intervals of time.

(T₄) A sortal-event can wholly occupy different intervals of time at different places, it being the whole event that occurs at each of these two different places.

By substituting 'sortal-event' for 'object' in definition (D$_t$) the following definition is derived:

(D$_t$)′ A sortal-event E wholly occupies an interval of time T at place S if, and only if, every temporal part of E occurring at S is contained in T and there is no sub-interval of T not filled at S by some temporal part of E.

There is no doubt that (S₄) is a conceptual truth, but is the same true for (T₄), when interpreted in terms of definition (D$_t$)′?

Far from being a conceptual truth, (T₄) is contradictory. Let us consider some sortal-event that satisfies the conditions of (T₄). Imagine a basketball game the first half of which is played on a court in place S₁ during interval of time T₁, and the second half of which is played on a different court in a different place S₄ during a later interval of time T₄. This game-event wholly occupies S₁ during T₁ and wholly occupies S₄ during T₄, thereby satisfying (T₄)'s conditions. However, it is only part of this game that occurs at S₁ during T₁, and likewise for S₄ during T₄. This should not surprise us since a sortal-event is *at least* made up of temporal parts. To say that the whole event occurs during each one of these temporal intervals (at these two different places) leads to an absurdity. If the whole game occurs during T₁ then the game begins at the beginning of T₁; and, similarly, if the whole game occurs during T₄ it begins at the beginning of T₄. But the beginning of T₄ is later than the beginning of T_1. Therefore, this event begins later than the time at which it begins!

Here is a further disanalogy between a sortal-object and sortal-event, which indicates that the sortal-nature of a sortal-event is not simply dependent upon temporal parts but also on the way it fills space, i.e. a sortal-event is both temporal and spatial. We have already considered this conceptual truth concerning a sortal-object's way of filling time:

(T₃) A sortal-object of a certain kind exists during any sub-interval of its history, and a sortal-object of that kind would exist during such a sub-interval even if the sortal-object were not to exist at any time earlier or later than that sub-interval.

The temporal analog, this time employing the concept of a sortal-event, is:

(S₃)' A sortal-event of a certain kind occurs in any sub-region of the space it occupies, and a sortal-event of that kind would occur in such a sub-region even if the sortal-event were not to occur at any space outside of that sub-region.

Once again, the spatial analog comes to grief upon the counterfactual clause. To see the conditions under which (S₃)' could be false we must indicate the manner in which an event occupies space. Events do take *place,* and, as a rule, the place in which they take place is the same as the place (s) occupied by the objects that directly participate in these events. A propagating event, such as a tidal wave, occupies the entire region of space over which it propagates. Let us work with a non-propagating sortal-event, such as a prizefight, which lasts for ten minutes. Assume that it consists in two fighters, A and B, standing stationary during this ten minute interval and throwing punches at each other. The space occupied by this event is the space occupied by both A and B. A sub-region of this space would be, e.g. the space occupied by A (or B). It seems clear that a sortal-event of this kind—a prizefight—does not occur within the space occupied by A alone, nor within that occupied by B alone. If the event were to consist only in what A does it would not be a prizefight, but rather an occurrence of shadow-boxing. Thus it is not true that a sortal-event of this sort—a prizefight—*would* occur in any sub-region of the space it occupies, even if this event *were* not to occupy any region of space outside this sub-region.

Because of these disanalogies between a sortal-object and sortal-event, $(T_2)'$ is not the temporal analog to (S_2). And it has already been argued at length that (T_2) is not a true analog either, since a sortal-object does not have temporal parts. While objects do not have temporal parts they do have histories. By a "temporal part" of an object we might understand some part of its history, i.e. some event or state in which the object is a direct participant. An object's history is made up of the various things it does and undergoes, as well as the different states it is in. We will consider only those events and states in which an object is a *direct* participant, i.e. is physically present during the time of the event or state and at the place at which it happens. To say that an object O wholly occupies an interval of time T at a place S, accordingly, will mean that every temporal part of O occurring at S occurs during T, and there is no sub-interval of T not filled at S by some temporal part of O. This means that every event and state occurring at S in which O is a direct participant occurs during T, and there is no sub-interval of T that is not filled by such an event or state. More simply put, O is at S throughout T and at no other time, this simplification being possible because an object cannot be at a place without participating in some event or state.

Utilizing this weak, disanalogous sense of the "temporal part" of an object, we can reformulate the temporal analog to (S_2) as follows:

$(T_2)''$ An object cannot wholly occupy both now and then at the same region of space.

This is a conceptual truth, because it means that it cannot be the case that an object occupies a given place now and at no other time and also occupies this place then, i.e. at some other time.

Even with this weak, disanalogous sense of 'temporal part' it can be shown that there are conceptual truths concerning 'now' that do not have spatial analogs. This is a conceptual truth concerning the use of 'now':

(T_5) If a person wholly occupies the same place S at two different and nonoverlapping intervals of time, T_1 and T_2, then the times denoted by his tokening of 'now' at these two different times are different.

'Now', as ordinarily used, denotes an interval of time that at least includes the time during which the sentence containing 'now' is tokened. The context determines the temporal width of the time denoted by a use of 'now'. We will assume throughout a principle of contextual constancy: if the first tokening of 'now' has a denotatum of n-time units then so has the later tokening of 'now'. The spatial analog to the above is:

(S$_5$) If a person wholly occupies the same interval of time T at two different and nonoverlapping places, S$_1$ and S$_2$, then the places denoted by his tokening of 'here' at these two different places are different.

'Here', as ordinarily used, denotes a place that at least includes the place occupied by the speaker. Once again, we assume the principle of contextual constancy: if one use of 'here' has a denotatum of n-cubic space units then so does the other.[8]

It can be shown that (S$_5$) could be false, and thus cannot be a true analog to conceptual truth (T$_5$). Imagine that spaces S$_1$ and S$_2$ are respectively the left and right hand parts of a person who remains stationary during T. Let us further suppose that during T this person simultaneously utters two different tokens of 'here', one being produced out of the left hand side of his mouth in S$_1$ and the other out of the right hand side of his mouth in S$_2$. Edward G. Robinson could probably accomplish this feat, but if the medical possibility of this worries you imagine that the person simultaneously holds up cards with 'here' inscribed on them, one held up by his left hand in S$_1$ and the other by his right hand in S$_2$. In either case there is the simultaneous tokening of 'here' in two different, non-overlapping places but nevertheless the same place referred to by each token. The reason is that a use of 'here' refers to a place which at least contains the place occupied by the speaker. Given the principle of contextual constancy, it is one and the same place denoted by these two uses of 'here', since they are used simultaneously by the same person. Thus, (S$_5$) is not a true analog to (T$_5$). This particular disanalogy lends further support to our earlier claim that objects have spatial but not temporal parts. It is reasonable to assume that if an object, such as a person, were made up of temporal as well as spatial parts, his tokenings of 'now' at different times at the same place should refer to the same time, just as his tokenings of 'here' at different places at the same time refer to the same place.

---

[8] It is not, of course, required by (S$_5$) that the person wholly occupy both S$_1$ and S$_2$ during T$_1$, which would be contradictory. The analogous contradiction would be for a person to wholly occupy two different times at the same place.

2. *Disanalogies Between 'Here' and 'Now' Concerning Action*

These disanalogies between 'here' and 'now' will be concerned with disanalogies between the way in which these terms enter into conceptual truths concerning what it is possible for an agent to do, which disanalogies bring out a sense in which 'here' but not 'now' is subjective. An agent can, within certain limits, choose where he shall live but not when he shall live. We should expect this dissimilarity between space and time to reveal itself in some conceptual disanalogy concerning the use of 'here' and 'now'. The following is a conceptual truth about an agent's power to alter his spatial position:

(S$_6$) A person who wholly occupies here now can bring it about that he wholly occupies there at a later time.

An agent who wholly occupies here now can deliberate about whether or not to move there, and thereby wholly occupy it at some later time. The temporal analog is:

(T$_6$) A person who wholly occupies now here can bring it about that he wholly occupies a later time there.

We took the liberty of substituting for 'there' in (S$_6$) 'a later time' in (T$_6$). The justification for this is that it is a conceptual impossibility for an agent to bring something about in the past.[9] It should now be clear that (T$_6$) is not a conceptual truth. For a person to bring it about that he will wholly occupy some later time T at a place S, he must bring it about that he occupies S throughout T and does not occupy S at any other time. But the 'at any other time' will include times earlier than T, *some of which are past.* Thus the agent must bring it about that in the past he did not occupy S. But to do this violates the conceptual truth that causation cannot work backwards from the present to the past.

There is a class of disanalogies between 'here' and 'now' having to do with there being numerous logical asymmetries between past and future concerning action that have no spatial analogs. For instance:

(T$_7$) An agent can now deliberate about what to do later than now but not earlier than now.

Because our present actions cannot bear past effects, we cannot deliberate about what to bring about in the past. The spatial analog to (T$_7$) is:

---

9 For a defense of this see Chapter Seven of my book, LT.

(S₇) An agent can now deliberate about what to do to the front (right, top) of here but not to the rear (left, bottom) of here. This seems to be absurd, and not just because space has three dimensions while time has only one. The point is that there is no logical asymmetry along any one of the three spatial dimensions. Thus, even if space had only one dimension this disanalogy would remain, although it is hard to imagine an agent existing in only one spatial dimension. Further disanalogies of this type are: An agent can now have intentions (choose in respect to) his conduct later than now but not earlier than now; to which the spatial analog is the absurd statement that an agent can now have intentions (choose in respect to) his conduct to the front (right, top) of here but not to the rear (left, bottom) of here.

It might be argued that I have failed to form the proper spatial analogs for these conceptual truths concerning 'now', since in the spatial analogs I retain the notion of deliberation, intention or choice, which are essentially temporal notions. I grant that these are essentially temporal notions, but what could their spatial analogs be? I cannot imagine what they could be. Without some such explanation this objection does not get off the ground.

A further disanalogy between 'here' and 'now', which shows that there is no spatial analog to temporal becoming, is due to there being no spatial analog to *waiting*. If I wait for *n*-time units my use of 'now' denotes a different time, regardless of whether I move or not. But the spatial analog to this is nonsense, since there is no sense to 'waiting for or through space'.

In conclusion, it seems to me that the disanalogies between 'here' and 'now' are profound: space and time are radically different.[10]

RICHARD M. GALE

UNIVERSITY OF PITTSBURGH

---

[10] I am deeply indebted to James Garson for stimulating me to work on this paper and for showing me how to deal with spatial and temporal analogies. Garson's criticism of my earlier work can be found in his article "Here and Now" in this volume.

STARTING AND STOPPING

At 8 a.m. I get in my car and set off for work. At 7:59 a.m., before I started it, my car was at rest; at 8:01 a.m. it is in motion. When a thing is not in motion, it is at rest, and when it is not at rest, it is in motion. But what was the state of the car *at* 8:00 a.m., as I was starting it? It would be inaccurate to say that it was in motion but it would be inaccurate, also, to say that it was at rest: it was "just starting." But, if whenever a thing is not in motion it is at rest, the state of "just starting"—or, of course, the comparable state of "just stopping"—must be either a state of motion or a state of rest. Which?

The problem is not one that is characteristic of changes of motion as distinct from other kinds of change. As the light comes on in a dark room, the room is apparently in some third state that is neither darkness nor nondarkness; and as I destroy a piece of paper it is apparently in a limbo between existence and nonexistence. Moreover, gradual changes raise the problem as much as sudden ones do; for either we think of the change as extending over a period, in which case the state of limbo is a prolonged one, or we stipulate a certain stage of the process as marking the transition, in which case the problem arises in respect of the time at which the process reaches this stage.

From a certain point of view this problem is a very trivial one indeed, and I certainly do not want to suggest that there is any mystery about *what happens* when changes take place. The interest of philosophers, since Parmenides and Zeno, in problems of this kind is misunderstood, sometimes, even by the philosophers themselves; but, come what may, it can hardly be suggested that the solution to our problem is to be sought by closer observation of accelerating vehicles or burning documents. It is not an empirical problem. Neither, on the other hand, is it one of how to describe the facts in ordinary English. I have just given a plausible description of them: the car is "just starting," and the light is "just coming on." In a paper entitled "The Origin of Motion"[1] this point is made at length by Brian Medlin, whose solution of the problem is to point out that to speak of a thing at time t as "at rest" or "in motion" normally suggests—though it does not entail—a period of uninterrupted rest or motion straddling t. Suggestions of this kind, I take Medlin to say, are carried around with most of our possible formulations of statements about change; but cause trouble only when we regard them as entailments. My feeling of unease in saying that my car is at rest at 8 a.m. is due to the impression that I am thereby involved in saying that it is uninterruptedly at rest throughout a period straddling 8 a.m.; but, if I need to be precise I can use a formulation that countermands this suggestion. Alternatively the matter could be resolved the other way, by countermanding the corresponding suggestion of the statement that the car is in motion. One way or another, the English language is equal to the task.

I agree with Medlin's solution of this part of the problem and it is not my task to criticise his paper. At the same time, it seems to me that there is a residual job to be done. Let us start by looking briefly at some of the other defences people have raised against the paradox. (Some of these are discussed by Medlin.)

(a) First, it might be objected that "starting a car" raises no problem because it is not an instantaneous process at all. "Starting my car" consists of getting in, adjusting my seat and doing my drill, and all sorts of other things. This is true but it is irrelevant. For present purposes the "time of starting" is the moment at which the clutch is finally engaged and the car moves.

(b) Secondly, however, it might be said that the example is unrealistic because even when the car does start it picks up speed gradually and not instantaneously. This seems to help us a little because it means that, at the moment power is applied, the instantaneous velocity of the car is still zero and it can be regarded as being at rest. On the other hand it could be argued that its average speed in any time-interval straddling the moment of starting is still non-zero, as before. In any case the problem is only pushed back, not solved, since the question that was asked of the velocity can be

---

[1] Brian Medlin, "The Origin of Motion," *Mind* 72, 286 (April, 1963), pp. 155-75. My attention was drawn to this paper by V. H. Dudman, to whom I am also indebted for a number of discussions.

asked of the acceleration. (If no derivative ever changes discontinuously, nothing ever changes.)

(c) Thirdly, it might be said: "The answer is arbitrary, because no *observation* is ever instantaneous, and hence any statement about the instantaneous value of a variable is without empirical meaning." This objection may be correct, but it opens up a whole range of related problems. In the first place, it is not clear whether one is supposed to reason that there is no such thing as an instant of time, or merely that no instant of time is ever observed separately from its fellows. The former course is the more heroic, but caution should be urged: there are difficulties in the concept of a time-scale without instants, for there must surely be time *intervals,* and intervals appear to require instants as their ends even if they are not, in fact, actually made up of continuous assemblages of them. The alternative view, that instants exist but are not separately observable, leaves us with the problem of saying what *is* observable. At t_0 do we, perhaps, observe instead of $P(t_o)$ some integral

$$\int_{t_1}^{t_2} P(t)dt$$

or weighted integral $\int_{-\infty}^{\infty} w(t\text{-}t_0)P(t)dt$?

The difficulty that attends any theory of integration over a finite range is the same as that which would replace time-instants by intervals: observations of events will have first and last moments, namely when the primary events of the underlying reality enter and leave the observation-window (t_1, t_2). On the other hand, an integral with infinite range and eternally non-zero weighting function permits influence on the present by very remote past and future events if these events are of a magnitude sufficient to compensate for their low weighting.

The *arbitrariness* of instantaneous values is itself disquieting. Before and after 8 a.m. the speed of my car was not arbitrary. (I am returning to the simpler hypothesis of instantaneous change of velocity.) In respect of 8 a.m. I may *either* say that its speed was zero, *or* that its speed was, say, 30 m.p.h., as I like; but this itself puts the matter of its speed at 8 a.m. in a very different light from that of its speed earlier and later. It is hardly stretching things too far to argue that this arbitrariness itself breaches the Law of the Excluded Middle.

It is worth while noticing that the thesis that time-instants must be replaced by intervals, rather than banishing three-valuedness for us, potentially opens the door wider to it; for intervals may have parts, and hence *no* predicate can be guaranteed to be uniformly true of, or false of, a whole interval. (Medlin says: "Not only does this argument fail to resolve the paradox. It needlessly extends it.") We shall look later at what needs to be done to the logic of intervals to meet this objection to it.

(d) "Time is unsharp"—this is what the wave-mechanical theorists say. It is a fundamental feature of physical reality that you can never achieve precision in one measurement, say a time-value, without sacrificing it in another, related one—in this case an energy-value. And, in order to succeed in measuring a time-value to a high degree of precision you must construct an experiment such that the energy-magnitude of the event concerned is highly uncertain, and the event will not be of a kind that it is of any use to you to observe. Of course, this effect is significant only in the case of microscopic events; but it illustrates, it would be said, the pointlessness of any attempt to be precise about instants.

It should be remarked that this objection, in spite of its different dress, is closely related to the previous one. Precise time-measurements must be replaced by "probability distributions," but it is not at all clear what it could mean to say that time-instants themselves *are* probability-distributions; and if, alternatively, the probabilities are epistemic ones, related difficulties arise as before.

(e) "The universe is noisy." This is another ploy of the physicist. My car, though it looks stationary, is really composed of molecules all of which are in rapid random motion, now left, now right. As I drive off, all that happens is that there is a long-term average movement of the swarm. The "moment of starting" is ill-defined.

This pushes the problem back into the micro-structure, but it otherwise does little to solve it. Individual molecules of the car must be sometimes changing from one state to another and hence have their instants of limbo. In any case, what concerns us is the motion, say, of the car's centre of gravity. The random motion of this centre will be very much less than that of individual molecules, and if we stipulate some small value of velocity as the boundary between "rest" and "motion" the car will change from rest to motion as the velocity of its centre increases through this value.

Micro-structure is, perhaps, of concern only to the physicist; but, if we confine ourselves to the macroscopic behaviour of the car, the problem reappears again. Either the car *does* instantaneously start and stop, in which case we are back where we were, or the indeterminacy of its near-starting behaviour is such that we must sum it up in a probability-distribution and face the other range of problems associated with objections (c) and (d).

Our problem is concerned with the *systematic* description of physical phenomena. It arises from the clash of the continuous-variable language of Mathematical Physics with the discrete two-valued language that we would like to make work as an alternative. The mathematical description appropriate to an accelerating object does not raise problems for us; but we would like, in addition, to be able to apply the two-valued predicate 'in motion', or its negation, to any thing at any time, and this is the project that runs into difficulties. If two-valued statements are predicates of time-instants they are *essentially* discontinuous: they cannot flow smoothly from falsehood to truth, and yet our time-language presents us with a continuum on which to define them.

It should be added that most everyday predicates are "durable." The red book on my desk could turn green for half a second or half a century but it could not turn green temporarily and durationless-ly at the stroke of twelve, remaining red at all times earlier and later. This being so, the time-continuum, modeled on the real numbers, is richer than we need for the modelling of empirical reality. Clicks, jerks, flashes, glimpses and impulses—which are our paradigms of "instantaneous" events—we would be well able to accommodate in a less lavish time-scale containing short but not infinitesimal intervals; and it begins to appear that *changes* of value of a predicate are the only true candidates for instantaneity. But, if this is so, why should we ever talk about truth or falsity *at* an instant? If instants exist at all, it is surely in some secondary sense derived from their role as assistants to intervals.

It does not seem at all impossible to build a logic in which intervals are the primary time-elements. That this would be a better logic for that part of language in which we use mainly two-valued durable predicates is, by now, obvious. That it may be less satisfac-

tory as a basis for the language of continuous variables need not
trouble us, since the concept of time as a continuum of instants has
itself run us into trouble with durable predicates; and, in any case,
we should suspend judgment until we see how satisfactory it can be
made to be. Let us build it.

The first and fundamental part of the task is to explore the
topological properties of *intervals*,[2] treated as entities in their own
right and not necessarily as sets of instants. Let us write '*a*,' '*b*', '*c*',
.... as variable names for them. The basic idea behind the logical
system I am about to describe is that intervals may *abut* one
another independently of whether there also exist other entities
describable as "points of abutment."

We can, in fact, use 'abutment' as our primitive notion in this
logic. I shall write '*aAb*' for '*a* abuts *b*', with the understanding
that this implies that *a* immediately precedes *b* so that '*bAa*' is not
equivalent to '*aAb*'. Taking abutment as primitive would enable
us to define, in general, '*a* precedes *b*' as '*Either a* abuts *b or* there
is another interval *c* such that *a* abuts *c* and *c* abuts *b*'. Allowing
ourselves the use of the lower predicate calculus, we can write this

$$a<b \equiv [aAb \lor (\exists c)(aAc \cdot cAb)].$$

For technical reasons it is actually easier to start with '$<$' as
primitive and set up a definition in the reverse direction.

We take as basis, then, the predicate calculus with the usual
truth-functional axioms and rules, the rules for quantifier introduction, the usual axioms for identity, and the single two-place predicate '$<$', to be read as 'is less than' or 'precedes'. At first we
shall need no predicate variables and can build a theory of intervals
entirely within the lower predicate calculus with identity. We define *abutment* by '*a* abuts *b* when *a* precedes *b* and there is no
interval such that *a* is less than *c* and *c* less than *b*'; that is

D1. $aAb = \mathrm{Df}\, a<b \cdot -(\exists c)(a<c \cdot c<b).$

We assume antireflexivity, antisymmetry and transitivity for '$<$':
actually, we can dispense with one or other of the first two and
deduce it as a theorem. Thus our first axioms are

---

[2] This is not to be confused with the "syntax of intervals" of Prior which is
worked out within a metrical logic of time; see A. N. Prior, *Past, Present and
Future* (Oxford, 1967), Chap. 5, hereinafter cited as PPF. I owe, however, a great
deal to Prior on this subject.

A1. $-(a<a)$ ("antireflexivity")
A2. $(a<b \cdot b<c) \supset a<c$ ("transitivity").

whence putting "a" for "c" in A2 and using A1 we deduce as our first theorem

T1. $a<b \supset -(b<a)$ ("antisymmetry")

However, we cannot assume all the usual properties of '$<$', since it is possible for two different intervals to be such that neither precedes the other: they may overlap. It is convenient to have a symbol for this, and we define 'aOb', to be read 'a overlaps b';

D2. $aOb = \mathrm{Df} -(a<b \vee b<a)$.

As we shall determine in due course, there are various different ways in which two intervals can overlap: one may contain another or not; and if a contains b it may be that they are both abutted by the same interval below, or that they both abut the same interval above, or neither. Since there are also two ways in which a may precede b, namely abutting or not abutting, and two ways in which it may be preceded by it, there are actually thirteen different possible relationships between a and b in terms of these concepts.

Abutment is antireflexive, antisymmetrical and antitransitive: overlapping is reflexive and symmetrical. In symbols these properties appear as

T2. $-aAa$
T3. $aAb \supset -bAa$
T4. $(aAb \cdot bAc) \supset -aAc$
T5. aOa
T6. $aOb \supset bOa$

Proofs of all these properties are direct, using the definitions A1, A2 and T1. T4 is proved as follows: From A1 we have $aAb \supset a<b$ and hence $(aAb \cdot bAc) \supset (a<b \cdot b<c)$; by A2, R.H.S. $\supset (\exists d)$ $(a<d \cdot d<c)$, and by D1 again, $\supset -aAc$. T5 illustrates that overlapping is compatible with actual identity.

There is nothing in the logic so far that positively differentiates it from a logic of points on a continuous line, since it is consistent with D1 that abutment should never actually occur, and consistent with D2 that overlapping should actually be the same thing as identity. We need to specify that when a and b do not abut there is some other interval which is abutted by one and abuts the other. Thus we have

A3. $a<b \supset [aAb \vee (\exists c)(aAc \cdot cAb)]$ ("connection")

whence the reciprocal "definition" of precedence in terms of abutment as suggested above is deducible. Provided only that there are
more than three intervals in the universe—for which we shall make
a stipulation in a moment—this axiom makes it definite that the
logic is not one of "points" in the usual sense. Thus if there are four
individuals a_1, a_2, a_3 and a_4, in order of precedence, so that a_1 abuts
a_2, a_2 abuts a_3 and a_3 abuts a_4, it must be the case by this axiom that
there is another individual a_5 which is abutted by a_1 and which
abuts a_4. That is, there must be an "individual" that combines some
properties of a_2 with some properties of a_3.

We are ready, in fact, to define *intersection* and *join* of intervals,
and *containment* of one interval by another. These come most
naturally in terms of the overlap function. We shall say that an
interval *a is contained in* an interval *b* if every interval that
overlaps *a* also overlaps *b*.

D.3.　$aCb = \mathrm{Df}\,(c)\,(cOa \supset cOb)$　("is contained in").

The intersection and join of two intervals, however, will not always
exist: the intersection will exist only if the two intervals overlap,
and the join only if they abut or overlap. (This means that we do
not have a "null interval" in our system. It would be possible to
make the logic work with a null element included, so that every
pair of intervals would have an intersection and join, but this
would complicate the definitions and axiomatisation in other respects, necessitating explicit exclusion of the null interval from
various formulae concerning abutment and order properties.) The
intersection of *a* and *b*, if it exists, is that interval that is overlapped
by just those intervals which overlap both *a* and *b*; and the join is
defined correspondingly with disjunction instead of conjunction in
the definiens:

D4.　$a \cap b = \mathrm{Df}\,(\imath c)\,(d)\,(dOc \equiv (dOa \cdot dOb))$　("intersection")
D5.　$a \cup b = \mathrm{Df}\,(\imath c)\,(d)\,(dOc \equiv (dOa \lor dOb))$　("join")

We need axioms guaranteeing their existence in appropriate cases;
thus:

A4.　$aOb \supset (\exists c)\,(d)\,(dOc \equiv (dOa \cdot dOb))$　("intersection")
A5.　$aAb \supset (\exists c)\,(d)\,(dOc \equiv (dOa \lor dOb))$　("join")

It is clear from the resemblance of the definitions to the usual ones
for Boolean concepts in terms of predicate calculus that *when the
intersection and join exist* the Boolean theorems relating them all

hold. In particular, without the formalities of numbering or proof, let us put down here for reference:

$$a \cap b = b \cap a$$
$$a \cap (b \cap c) = (a \cap b) \cap c$$
$$a \cap a = a$$
$$a \cup b = b \cup a$$
$$a \cup (b \cup c) = (a \cup b) \cup c$$
$$a \cup a = a$$
$$a \cup (b \cap c) = (a \cup b) \cap (a \cup c)$$
$$a \cap (b \cup c) = (a \cap b) \cup (a \cap c)$$

To be valid theorems, some of these formulae need to be prefixed with conditions adequate to guarantee the existence of the various intersections and joins. (In one case, the last, we may otherwise even have one side of the identity existing without the other.)

Intervals thus have a combination of Boolean and ordering properties. Postponing further theorems, let us complete our axiom system. At this stage we have a fairly open choice of what to build in. There are two particular respects in which this is so, namely (1) the question of whether intervals are to be regarded as infinitely divisible, and (2) the behaviour of intervals as regards ultimate past and future. I shall, for the moment, assume infinite divisibility. An axiom sufficient in context to guarantee this is:

A6. $(\exists b) (b C a \cdot b \neq a)$ ("divisibility").

This says that every interval properly contains another; and it will follow, with the use of A3, that every interval can be subdivided into at least two parts. This means that we are, at present, essentially building a logic of a "dense" time-scale, and that it is consistent with a time-scale made up of intervals that it should be so.

On the question of ultimate past and future it would not be unreasonable for us to suppose that there exists an "ultimate future interval" which does not abut or precede any other; and similarly for the past. These assumptions are not even inconsistent with the concept of an infinite past and future as we normally understand them, since there is no a priori objection to the notion of an infinite interval. We could secure this result by the addition of a single axiom of the form 'There exists an interval which overlaps every other', or:

$$(\exists a) (b) (a O b) \quad \text{("universe").}$$

This "universal interval" could be given a name and play the role of a universal class in the Boolean formulae. Nevertheless the assumption of its existence is not obligatory, and we can easily suppose that every interval is preceded and followed by others. This is a convenient present assumption, and we accordingly have:

A7.1 $(\exists b)\,(a<b)$ ("infinite future")
A7.2 $(\exists b)\,(b<a)$ ("infinite past").

Alternatively we could specify the existence of an ultimate interval in the past but not in the future, or *vice versa*. It should be noted that with the choice of axioms we have actually made there is complete symmetry between past and future, so that it is, in fact, arbitrary which is which.

Now let us develop some of the further consequences of the axioms. Two theorems of importance in deriving others are those which specify that when a abuts b and b abuts c, a abuts the join of b and c, and the join of a and b abuts c:

T7.1 $(aAb \cdot bAc) \supset aAb \cup c$
T7.2 $(aAb \cdot bAc) \supset a \cup bAc$

The proof of T7.1 is as follows: The existence of $b \cup c$ is guaranteed from bAc by A5. If a does not abut $b \cup c$ there are three possibilities: (1) $(\exists d)\,(aAd \cdot dAb \cup c)$. Then $a<d$, $d<b \cup c$; but we easily prove that it is impossible that $d<b$ or that $d<c$, whence $b<d$ and $c<d$, and $b \cup c$ cannot possibly satisfy its definition. (2) $aOb \cup c$. Then $aOb \lor aOc$ by the definition, and we cannot have $aAb \cdot aAc$. (3) $b \cup c<a$. This is inconsistent with the antecedent by reasoning similar to (1).

From this we prove

T8.1 $(aAb \cdot aAc) \supset bOc$
T8.2 $(aAb \cdot aAc) \supset aOb$

again by dividing cases. (Thus for T8.1 it is sufficient to prove inconsistency of $aAb \cdot aAc \cdot b<c$, and the case $b<c$ may be divided into the cases bAc and $bAd \cdot dAc$.) Similarly we have

T9.1 $(aAb \cdot aAc) \supset aAb \cap c$
T9.2 $(aAc \cdot bAc) \supset a \cap bAc$
T10.1 $(aAb \cdot aAc) \supset aAb \cup c$
T10.2 $(aAc \cdot bAc) \supset a \cup bAc$

Containment is Boolean in its properties. Thus if the intersection exists we have

$aCb \equiv (a \cap b = a)$.

By subdivision of cases we prove that two intervals are identical if they overlap the same intervals:

T11. $(c)\,(cOa \equiv cOb) \supset a = b$

and hence, from the definition of containment, if they contain one another; that is:

T12. $(aCb \cdot bCa) \supset a = b.$

Although there is no universal interval in the sense of an interval which overlaps all others, we can easily construct an interval overlapping any finite number of given ones. Thus, given a and b, we have

T13. $(\exists c)\,(cOa \cdot cOb).$

This is immediate for aOb: if $a < b$ or $b < a$ we can construct their join, if they abut, or their join together with an interval connecting them (by A3). The axioms A7.1 and A7.2 that specify the non-existence of ultimate future and past intervals can easily (also using A3) be strengthened to specify the existence of an interval abutting, and an interval abutted by, any given interval; thus:

T14.1 $(\exists b)\,(aAb)$

T14.2 $(\exists b)\,(bAa).$

What might be called the fundamental theorem of density says that any interval may be divided into a pair of *abutting* intervals:

T15. $(\exists b)\,(\exists c)\,(bAc \cdot a = b \cup c).$

This is proved by first postulating an interval contained in a, by A6, and then, if necessary, postulating other intervals to fill the gaps at the end of the range, by A3; after which the three intervals that result in the most general case can be reduced to two by joining.

It remains to notice that A5 did not give the most general conditions for existence of join. We can prove:

T16. $(\exists c)\,(c = a \cup b) \supset (aAb \vee bAa \vee aOb).$

Since the system has been built entirely within the lower predicate calculus it follows from known results that a model of the system can be built on, at most, an enumerable infinity of individuals; that is, of intervals. It is clear, moreover, in view of the "infinite divisibility" theorem and the "infinite past" and "infinite future" axioms, that only a very artificial model would make do with fewer. A consistency proof relative to the algebra of rational numbers can be given very simply by representing intervals as ordered pairs of numbers with the first member of each pair less than the second: abutment of a to b is equality of the second member of a to the first member of b, and so on. Consequently there is nothing in the system that is far out-of-line with orthodox representations of time. Only the absence of "instants" is in any way unusual.

Let us now see how the logic of two-valued tensed statements squares with this conception of time. Let each statement be treated as a predicate of a time-variable: thus if 'p' is such a statement let it be treated as incomplete and as being required to occur in the context 'pa' ('p is true at time a'), for some a. We are adopting the principle that statements—which we take as stating simple facts such as that a certain thing is red, or that the room is dark, or that the car is at rest—are true not at instants but during intervals. It follows that the "statements" we now introduce will be predicates taking intervals as values. Consequently, at this point, we move beyond the lower predicate calculus.

Yet it will immediately be seen that we cannot get by, at this stage, with two-valued predicates. During an interval, there are at least three possibilities with regard to an elementary statement: it can be true throughout the interval, false throughout the interval, or at times true and at times false. It is possible, that is, that p may be true in all subintervals of a, or false in all subintervals; but it may also be the case that there is at least one subinterval in which it is true and at least one in which it is false. Even so, there may be some subintervals in which it is neither true nor false. We cannot, at this stage, avoid three-valuedness.

Grasping the nettle, let us write 'p^+a' for 'p is true throughout a', 'p^-a' for 'p is false throughout a' and 'p^*a' for 'p is sometimes true and sometimes false in a', that is, 'p changes in a'. These statements will be treated as a trio of contraries, so that the negation of any one of them is equivalent to the disjunction of the other two. (In particular, note that p^-a is not equivalent to $-(p^+a)$.) We now have two kinds of specification to lay down. (I shall not attempt an axiomatisation of this three-valued system.) First, there will be rules relating the truth of p in a to its truth in other intervals topologically related to a; and secondly, there will be rules relating the truth of p in a to the truth in a of other statements logically related to p. The first part of this task is easy; but it will be necessary to treat predicates of types 'p^+' and 'p^-' differently from type 'p^*'. When 'p^+' or 'p^-' is true of a given interval it will also be true of any interval contained in it; thus:

$$bCa \supset (p^+a \supset p^+b)$$
$$bCa \supset (p^-a \supset p^-b).$$

For 'p^*', the reverse is the case; if it is true of a it is true of any interval that *contains* a:

$$aCb \supset (p*a \supset p*b).$$

One result of this asymmetry is that we cannot freely substitute a '$p*$' for a 'p^+' or 'p^-', or *vice versa*, in deriving theorems within this logic; nor can we freely substitute, say '$-p^+$' for 'p^+'.

In the second part of our task we strike a similar awkwardness. Given, say, p^+a and q^+a we can cheerfully write their conjunction in some such form as '$(p^+ \cdot q^+)a$' or even '$(p \cdot q)^+a$'; but we cannot do the same with $p*a$ and $q*a$, since changes of truth-value in p and q might take place together in such a way as to compensate one another. The same will apply with other truth-functional operators. A three-valued logic is in fact an embarrassment, and we would do well to do something to get rid of it.

Before we do so, however, let me point out an interesting application of the system to "modal"-type tense-logics. Let 'F', 'P', 'G' and 'H' represent the four propositional operators 'It will at some time be true that', 'It was at some time true that', 'It will always be true that' and 'It has always been true that'. The range of logical systems that can be built has been thoroughly explored by Prior.[3] One way of modelling the properties of the tense-operators is to treat statements $p, q, r, \ldots$ as predicates of time-instants, to be written $pt, qt, rt, \ldots$, whence, if we choose a particular instant n as "now" we can correlate

'Fp' with '$(\exists t)(t>n \cdot pt)$',
'Pp' with '$(\exists t)(t<n \cdot pt)$',
'Gp' with '$(t)(t>n \supset pt)$', and
'Hp' with '$(t)(t<n \supset pt)$'.

If we treat the time-variable t as ranging over intervals instead of instants and (uncritically) identify the 'p' in each formula on the right with our 'p^+', we can take over these as definitions and will get a new tense-modal logic in F, P, G and H. Prior has given various axiom-sets appropriate to "discrete" and "dense" time-scales respectively, and we would expect our system to yield some combination of these. It turns out to contain all of a set of formulae I suggested, rather brashly, as axioms for tense-logic in 1957. My set of axioms was not published at the time but has been commented on in the literature[4]: it has generally been regarded as a very peculiar system

3 PPF, Chaps. 3-6.

4 Prior, PPF; see pp. 47-50.

which, indeed, it is. One of the formulae, $Fp \supset FFp$, states in effect that when p is going to be true there is a future time between the present and the time of p; that is, a time at which it will be true that p will be true. This will be the case, in the ordinary way, if the time-scale is "dense" or "continuous" but not if it is "discrete," since in the latter instance there may be a "next future instant" at which p is true. Another of my formulae, $(p \cdot Hp) \supset FHp$, has the precisely opposite implication since it states that if p is true now and always has been in the past there is a future time at which it will always have been true; and this can only be true in general if there *is* a "next future instant." The other axioms of my set all also come out as valid. I should add that no one could be more surprised by this result than I was myself.

When we review the implications of the three-valuedness of our propositions the result recedes a little: we do not have a true interpretation of my axiom set, since we cannot freely use the substitution rule in deriving theorems. One other formula, $PGp \supset Pp$, that has sometimes been taken as characteristic of a "dense" time-scale, actually comes out as false. Since this is provable in the original system we seem to have a contradiction; but it happens that a substitution on which the proof depends—the putting of Gp for p in $Pp \supset PPp$—is not valid under the identifications proposed.

What can we do about the three-valuedness of the statements which are predicates of intervals? If we could subdivide intervals in some way which would lead us eventually to "atoms," within any one of which no change takes place, we would have solved the problem; but is the world really such that this is possible? The doctrine traditionally attributed to Heraclitus—that everything is always changing—is to the effect that it is not possible. Yet it so happens that it is very reasonable at least to explore the construction of a theory in which certain time-intervals have a privileged position. Heraclitus, we would suppose, thought of "change" as a continuous, flowing process; but, if what is changing, linguistically expressed, is the truth-value of temporal predicates, it is difficult to imagine that changes are anything but discrete and spaced. The everywhere-discontinuous functions of Pure Mathematics do not seem to represent elementary empirical processes at all. And, in any case, we must remind ourselves that even infinite subdivision does not necessarily give us entities with the properties of instants, incapable of division any further.

Let us suppose that our language contains elementary statements $P_1, P_2, P_3, \ldots$, finite or infinite in number, and imagine the universe as a time-sequence of states describable by conjunction-chains of these statements, straight or negated. We could, perhaps, specify that no state is ever repeated. In any case, each state can be predicated of some interval of time, and the intervals of time so singled out will abut one another and will form a "natural" set of time-elements. We need not suppose either that they are indivisible or that they cannot be joined up into larger intervals; but they will play a special role in the theory.

An *elementary interval* is an interval within which no change of truth-value of any of the P_i occurs, and such that it cannot be enlarged without including such a change. That is to say, intervals which overlap it contain a truth-value change if and only if they are not contained in it. In symbols:

$$Ea = \mathrm{Df}\ (b)\ \{bOa \supset [bCa \equiv (p)\ (-p*b)]\}$$

Here it is to be understood that the predicate-variable P ranges over the "empirical" predicates $P_1, P_2, P_3, \ldots$ and their truth-functions, and not over predicates representing topological properties of the time-scale (such as the predicate of a represented by bOa). In what follows I shall use Greek letters $a, \beta, \ldots$ for elementary intervals and omit any explicit reference to the above definition, writing, for example '$(\exists a)\ (\text{-}\text{-}\text{-})$' for '$(\exists a)\ (Ea.\text{-}\text{-}\text{-})$'. Now for any elementary equivalent to $-p^+a$, so that we can omit '+' and '-' and treat our predicates as two-valued. Moreover, we can dispense with three-valuedness in the case of *other* intervals also, except in so far as it enters by way of the definition of "elementary interval"; for, for any interval a whatever,

$$p*a \equiv (\exists a)\ (aOa \cdot pa) \cdot (\exists \beta)\ (\beta Oa \cdot -p\beta)$$

and similarly

$$P^+a \equiv (a)\ (aOa \supset pa)$$
$$p^-a \equiv (a)\ (aOa \supset -pa).$$

The logic of *elementary* intervals is the logic of discrete time-instants, finite or infinite in number. Its interest, however, is as a basis for the logic of intervals in general. If, instead of introducing elementarity by definition, we treat it as a new primitive, the logic of intervals becomes fully two-valued and we have a complete basis for a logic of time without instants. If we want to model, in this logic, some statement about an instant of truth-value change of a durable predicate—and there is no *other* kind of instant that we need to make room for—it will be as a statement about the respective values of that predicate in a pair of abutting elementary intervals. In some respects this analysis is in the spirit of Dedekind, and treats the rational numbers as he did the reals.

If the usefulness of real numbers in Physics is such as to encourage their construction even on this foreign basis this can, perhaps be done.

I take my title from William Heytesbury, of Merton College, Oxford, who discussed a family of related problems in his *Regule solvendi sophismata,* in a chapter entitled "De incipit et desinit," about the year 1335.[5,6]

C. L. HAMBLIN

UNIVERSITY OF NEW SOUTH WALES

------

[5] Heytesbury and his colleagues enlarged the mediaeval explications of the terms *incipit* and *desinit* into a broader theory of bounds and limits. See Curtis Wilson, *William Heytesbury: Medieval Logic and the Rise of Mathematical Physics*, Madison, Wis., 1956.

[6] (Note added in proof) : Since writing the above I have discovered that the given set of axioms is incomplete for the logic of intervals, affecting particularly the deduction of T11 and T12. A repaired set is presented and discussed in my paper "Instants and Intervals," *Studium Generale,* 27 (1971), 127-34. The main conclusions of this article are unaffected by the error.

TIME AND THE PHYSICAL MODALITIES

Relative to any point in time, how many possible futures are there? For example, it may rain tomorrow, or again it may not. So it would appear that relative to today, there are at least two possible futures, one involving rain tomorrow and the other not. Of course only one of these two future states of affairs will take place, and in that sense there is only one *actual* future, though there may be many *possible* futures. The only hypothesis under which there is, for every instant in time, only one possible future, is the hypothesis of universal Laplacean determinism, and this hypothesis has had little to recommend it since the advent of quantum physics. Furthermore, although among the many possible futures relative to any given moment in time there is only one which will become actual, no distinguishing mark separates this one from its fellows *at the given moment*. So it would appear that what confronts the world at each moment in time is an undifferentiated multiplicity of possible futures. But this multiplicity, though at first sight admittedly frightening in its implications, may be put to good philosophical use: as will be seen it yields natural and convincing definitions of the notions of physical necessity and possibility. On the other hand, alternative definitions of these notions are obtainable which do not make use of the manifold of possible futures, and these also will be presented. The paper will be in two parts, the first dealing with the concepts of physical possibility and necessity, and the second presenting a formal calculus of necessary and sufficient conditions in which the physical modalities find their place.

I
Physical Possibility and Necessity
1. *Relative Modalities*

Conceiving of a manifold of alternative futures will throw considerable light upon the notions of physical possibility and necessity, or, as is sometimes said, causal possibility and necessity.[1] Hitherto a fair amount has been written about these modalities under the hypothesis that they are either identical with or resemble the logical modalities in most important ways. But there is one supremely important respect in which they differ, and once this difference is recognized a number of confusions disappear. The difference may be put in this way: whatever is physically possible or necessary is possible or necessary *relative to a certain state of affairs,* or set of initial conditions at a certain time. What is logically possible or necessary, on the other hand, is so unconditionally, without reference to anything except perhaps the laws of logic. For example, it is logically possible that I should drop my pen, and that instead of falling it should rise into the air. Or that it should explode or vanish or turn into the Taj Mahal or anything which did not imply a contradiction or otherwise do violence to logic. All these things are logically possible in a quite unconditioned way, as is also the case with the logical necessity of my pen's *not* turning into a round square, or becoming red and green all over. No reference to antecedent conditions is needed to determine what is logically possible in the future history of my pen. Some things are logically possible, and that is all.

With physical possibility and necessity, however, the matter is quite otherwise. If, sitting at my desk, I drop my pen, it is physically impossible that it should rise into the air, and it is physically necessary that it should hit the table-top. But wait, someone will say, suppose a sudden wind arose, or a bird swooped in at the window and seized the pen before it hit the table? Then I should take care to shut the window, and would drop the pen holding it only a fraction of an inch above the table. If someone were to object that the air molecules beneath the pen might miraculously find themselves in a 'solid' configuration, so that the air froze, I should perform the experiment in a vacuum. Eventually I should arrive at a future state of affairs that was physically or causally necessary, relative to a set of initial conditions, but not logically necessary. Note that it makes no sense to speak of *my pen hitting the table* as

---

1 For a discussion of the causal modalities see Wilfrid Sellars, "Counterfactuals, Dispositions, and the Causal Modalities," *Minnesota Studies in the Philosophy of Science* Vol. II, 1957. The term 'physical modality' stems from Reichenbach (New York: *Elements of Symbolic Logic,* 1947) , p. 392.

physically necessary, but only *my pen hitting the table relative to my dropping it in a certain way, and under certain conditions.*

Failure to observe the improperness of saying '*A* is physically possible', rather than '*A* is physically possible relative to *B*', has led to many fruitless arguments. Is it possible for the sun not to rise tomorrow? Doubtless no, relative to the current state of the solar system, but if a comet were approaching the earth, whose impact would stop its rotation, then doubtless yes. Must bread nourish, or could it poison a man? Well, if the bread's initial state includes the presence of ergot it will certainly not nourish. But an unconditional answer to these puzzles, without mentioning the initial state of affairs to which the possibility or necessity is referred, is plainly out of the question. The physical modalities are inherently relative, not absolute.

At the risk of tedium, let us once more consider the distinction between the causal and the logical modalities by examining the question, is there anything which is logically possible but physically impossible? For if not, the two concepts will be at least extensionally equivalent. Now it is true that a great many things which up to a century ago were thought to be physically impossible are now regarded as possible. If I am sitting in a bath (to borrow an example from Professor Grünbaum) then it is possible, though exceedingly unlikely, that the high-energy and low-energy water molecules should separate, so that my head boils and my feet freeze in a block of ice. A baseball, thrown at a wall, will pass right through in the improbable event that its molecules fit exactly into the wall's interstices. But we should not conclude from these examples that it is physically possible for *anything* to happen. On the contrary, given appropriate initial conditions many things are not only improbable but actually impossible. For example it is physically impossible for the moon, circling the earth on a summer's night in a clear and unobstructed sky, suddenly to reverse itself and circle in the opposite direction. Or, if there might still be thought to be some taint of statistical probability attached to the motion of such a large collection of particles as the moon, consider a single particle, say a proton, travelling through empty space. It would be physically perfectly impossible, given initial conditions which included no change in the gravitational or electromagnetic field, for this proton to reverse its motion. It would however be logically possible for it to do so, since the violation of the law of inertia which such a reversal implies is not the violation of a law of logic. This example shows that the concepts of physical and of logical possibility are not extensionally equivalent. But it also indicates that if physical possibility were not a relative notion they *would* be extensionally equivalent, since it is physically quite possible (in an absolute sense) for a proton to reverse its motion.[2]

2. *Leibnizian Definition of the Physical Modalities*

I now wish to turn to the manifold of alternative futures. The connexion between this manifold and the physical modalities is the plain and obvious Leibnizian one: what is physically necessary is what holds in all possible futures, and what is physically possible is what holds in at least one. If they are to be precise, statements involving the causal modalities must contain a double time reference.[3] For example, if I drop my pen at time t_1, then it is necessary that it hit the table at time t_2 if and only if all possible futures relative to the time t_1 exhibit the pen hitting the table at time t_2. If one single possible future relative to t_1 does *not* exhibit the pen hitting the table at t_2, then it is possible for it not to hit, hence not necessary for it to hit. In this way the causal modalities are definable in terms of the manifold of possible futures. The full meaning of 'A is physically necessary' is 'A at t_2 is necessary at t_1'; that is, according to this interpretation, 'A at t_2 occurs in all possible futures relative to t_1'.

Notice now that nothing prevents us identifying the two times in the above paragraph and taking 'A at t_1 is necessary at t_1' to mean 'A at t_1 occurs in all possible futures relative to t_1', as long as 'future' includes all moments simultaneous with or later than the present. We have as a consequence Aristotle's well-known but somewhat puzzling doctrine that 'whatever is, *when* it is, is necessary; and whatever is not, *when* it is not, necessarily is not'. This doctrine is incoherent on the supposition that any other than relative necessity is in question. Thus, *relative* to the salt-cellar's being in the middle of the table at time t_1, it is necessary that it be in the

---

[2] I.e., in a sufficiently strong electrical field.

[3] This phrase comes from K. Lehrer and R. Taylor, "Time, Truth and Modalities," *Mind* 74 (1965), p. 395.

middle of the table at t_1 and nowhere else. It follows that it is impossible that it be anywhere else, such as on the side of the table. Once it is perceived that this latter impossibility is relative to the time t_1, the air of mystery surrounding the modality in question is dispelled, since the obvious objection that the salt-cellar *could* be somewhere else (or, more correctly, *could have been* somewhere else) is answered. Relative to the time t_0, perhaps, the salt-cellar could have been on the chair at time t_1 (meaning that it was possible at t_0 that the cellar should be on the chair at t_1), but relative to t_1 the cellar could be nowhere else than where it is at t_1, namely in the middle of the table. It follows that certain modal fallacies which attend the logical modalities lose their sting when applied to the physical modalities, as will be seen.

In any logic book, the following inferences will be invalid:

It is necessary that, if p, then q

p

Therefore, it is necessary that q.

(This inference involves passing from a *necessitas consequentiae* to a *necessitas consequentis*.)

It is impossible that both p and q

p

Therefore, it is impossible that q.

But these inferences are not invalid when the modalities in question are physical and relative. It is necessary that if the salt-cellar is on the table at t_1 it is not on the chair at t_1; it is on the table at t_1; therefore, it is necessary that it is not on the chair at t_1 (it being understood that this necessity is relative to t_1). It is impossible for me both to walk to work and to take the bus to work; I walk to work; therefore it is impossible for me to take the bus. It is doubtless examples like these, which seem to violate recognized principles of modality, that have made it so difficult for philosophers to understand and accept Aristotle's doctrine of necessity in *De interp.* *ix*. But, seen in the right light, the examples are perfectly harmless.

One further consequence of what has been said about the physical modalities and their relationship with the manifold of possible futures is that it may now be seen in what sense the past is necessary. Whatever happens in the present is necessary, relative to the present, because although there may be many possible futures there is only one present. Similarly whatever has happened in the past is necessary, relative to the present, because there is only one past. As of 1968 it is not possible that World War II continued into 1946, although as of 1944 it was possible. In this sense the past and the present are necessary, and in this sense even the most chancy event, such as a fall of a coin, becomes necessary once it occurs. For this reason the type of necessity that Aristotle talks of in *De interp. ix* is sometimes called *temporal* necessity, to distinguish it from causal necessity which requires the operation of antecedent causes. But in fact temporal and causal necessity, looked at from the Leibnizian standpoint, are not so different. Both are relative modalities, not absolute, and both define the necessary as that to which no possible alternatives exist (in the case of temporal necessity this requirement is vacuously satisfied by the uniqueness of the past). There is, however, another approach to the physical modalities which differs from the Leibnizian one, and to this we shall now turn.

3. *Necessary and Sufficient Conditions*

The physical modalities were defined above in terms of a multiplicity of possible futures, coupled with a unique present and past. But there is another way of defining these modalities, which consists in replacing the multiplicity of possible futures by the concept of necessary and sufficient conditions. This mode of definition proceeds as follows. A at time t_2 is physically necessary relative to time t_1 if and only if there exists at t_1 some condition sufficient for A at t_2. Again, A at t_2 is physically possible relative to t_1 if and only if there exists at t_1 no condition sufficient for the absence or nonoccurrence of A at t_2. Note that, as before, physical necessity and possibility are relative notions. An example will illustrate these definitions.

If I pull the trigger of a gun at time t_1 then, assuming that the gun is loaded, that the safety catch is not on, etc., there exists at t_1 a state of affairs sufficient for the exit of the bullet from the muzzle at time t_2. That is, the exit of the bullet at t_2 is physically necessary relative to t_1. Furthermore, if at t_2 there is not a strong wind blowing, if the gun's sights are correctly adjusted, if my aim is good, and if I squeeze the trigger smoothly, there exists at t_2 no condition sufficient to prevent the bullet hitting a target 1000 yards away at time t_3. That is to say, the bullet's hitting the target at t_3 is physically possible relative to t_2. If we wish, we may further define the notion of *contingency* as follows. A at t_2 is contingent relative to t_1 if both A and the absence of A are possible relative to t_1.

Then in the case of the gun it is doubtless contingent, relative to the time t_2, that the bullet should hit the very centre of the target at time t_3.

The results of defining the physical modalities in this way would seem to coincide exactly with the results of defining them in terms of possible futures, as long as it is the modality of future or present events that is in question. Thus if there exists at t_1 a condition sufficient for A at t_2, then A will be in all possible futures relative to t_1, and it would seem reasonable to think that the converse holds as well. Similarly in the case where A is possible. If $t_1 = t_2$, we may stipulate that A be a sufficient condition of itself. But it does not seem plausible to assert that every past event is necessary relative to the present, if this means that there currently exists a sufficient condition for every past event. No doubt for some past events this is true. Thus if buying a ticket is necessary for entering a movie theatre, entering a movie theatre is sufficient for having bought a ticket. But it is very difficult to think what might now be sufficient for Brutus having killed Caesar.[4] Hence if necessity is defined in terms of sufficient conditions, it is dubious that every past event is necessary (though certainly true that every past event is possible).

The definition of the physical modalities in terms of necessary and sufficient conditions is a powerful one, and is able to embrace and render plausible a species of impossibility which has recently generated a fair amount of controversy. Consider the following principle:

(i) Nothing can take place, a necessary condition of which is lacking.

This principle is a generalization of one concerning human action which has been proposed by Richard Taylor:

(ii) No one can perform any action, a necessary condition of which is lacking.

It will be argued, using examples, that when the proviso is made that the missing necessary condition may be either past or present, but not future, principles (i) and (ii) are true.[5] It will then be

---

[4] Unless of course we allow states of affairs such as 'tomorrow's being the anniversary of Caesar's death at the hands of Brutus'. We need to exclude these "temporally impure" descriptions—see Richard Gale, *The Language of Time* (London, 1968), pp. 155-64.

[5] Principle (ii) makes its first appearance in R. Taylor, "Fatalism," *The Philosophical Review* 71 (1962), p. 58. The two different versions of the principle

shown how these principles may be derived from the definition of necessity in terms of necessary and sufficient conditions, together with two other principles of high plausibility.

To support the truth of principle (ii) the earlier example of the entry ticket will suffice, except that this time it will be a ticket to a swimming pool. If having a ticket is a necessary condition of entering, and if at time t_1 I have no ticket, then at time t_1 it is impossible for me to enter. This is an example of a present necessary condition which is lacking; for a past necessary condition we may suppose that on crowded days it is necessary to have purchased a ticket by 4 p. m. in order to enter at 5 p. m. If I fail to purchase a ticket by 4 p. m., then, relative to 4:30 p. m., I cannot enter at 5:00 p. m. (Note that there is a *triple* time reference involved here.) What now of the case of a missing future necessary condition? Let us suppose that I have successfully entered the building, and am standing on the edge of the pool at 7 p. m. Let us suppose too that the other swimmers have left, and that the water is perfectly calm. Now (taking an example from Saunders) my swimming at 7 p. m. is sufficient for the water to be turbulent at 7:01 p. m., so the water's being turbulent at 7:01 is necessary for my swimming at 7. But suppose that the water is not turbulent at 7:01, and that consequently a future necessary condition of my swimming is lacking. Does this mean I cannot swim at 7? Not at all. It means only that I *do* not swim. I manifestly *can* swim at 7, since I have the ability, the opportunity, etc. But perhaps I do not choose to.

These examples indicate that principle (ii) is true, and it is easy to generalize them so as to apply to principle (i). For this, one has only to remove the human element. Suppose an engineer is showing a visitor around a chemical plant, and points out a pipe through which oxygen is flowing. The visitor asks why the oxygen doesn't reverse its flow, backing up into a tank of acetylene. The engineer

---

which result from distinguishing between missing past and future necessary conditions are introduced by John Turk Saunders in "Fatalism and Ordinary Language," *The Journal of Philosophy* 62 (1965), p. 213. For a discussion of the consequences of this distinction for Taylor's arguments concerning fatalism, see the present author's review of Steven Cahn's "Fate, Logic and Time," *The Journal of Philosophy* 65 No. 22, (1968), pp. 742-46, and, for an analysis of human ability in terms of physical possibility, "Ability as a Species of Possibility," forthcoming in *Human Action*, edited by Myles Brand.

says it can't, because the pipe has a one-way valve attached to it. Note that the engineer does not say merely that the oxygen *doesn't* flow backward, but that it is impossible for it to do so. And he is quite justified in saying this: the engineer and the plant manager can sleep happily, secure in the knowledge that because of the one-way valve, the oxygen cannot mix with the acetylene. The reason is plain; a necessary condition of their mixing is lacking. Similar examples will show that nothing can happen, a past necessary condition of which is lacking, and that things *can* happen even though they lack future necessary conditions.

What kind of impossibility is reflected in principles (i) and (ii)? To say that A cannot happen because one of its necessary conditions is lacking is no far cry from saying that A cannot happen because there exists a state of affairs sufficient for its nonoccurrence. Hence if we adopt the following principle:

(iii) If a necessary condition for A is lacking, then a sufficient condition for the nonoccurrence of A is present

then principles (i) and (ii) are derivable from the definition of necessity and impossibility in terms of necessary and sufficient conditions. It remains to show why a missing past or present necessary condition entails the impossibility in question, whereas a missing future necessary condition does not.

If a necessary condition of A is lacking at time t_1 then A is impossible relative to t_1 by principle (iii) and the definition of impossibility. Suppose now that a *past* necessary condition of A were lacking at t_1; i.e. that some state of affairs B at time t_0 were necessary for A, and that B did not occur at t_0. Then by principle (iii) there existed at t_0 a state of affairs sufficient for the nonoccurrence of A, and A would be impossible relative to t_0. To show that A would also be impossible relative to t_1, a further principle is necessary:

(iv) If there exists at t_0 a sufficient condition for A at t_1, then there exists at all times between t_0 and t_1 a sufficient condition for A at t_1.

We might call principle (iv) the *principle of the continuity of sufficient conditions*. To take an example, if at t_0 a valve in a pipe is opened which ensures the exit of water from the end of the pipe at t_1, then at all moments between t_0 and t_1 there exists a sufficient condition for the exit of water at t_1. (Note that this sufficient condition need not be the *same* one as existed at t_0, which includes the valve being open, since the valve could be closed again and the water still exit at t_1. All that is needed is some sufficient condition or other at every instant between t_0 and t_1.) It is of interest to note that what is necessary relative to an earlier time cannot cease to be necessary until the moment it occurs, whereas what is possible relative to an earlier time can cease to be possible at any intermediate time.[6]

Finally, why does a missing future necessary condition of A not imply that A is impossible? Suppose that $t_1 < t_2 < t_3$, and that at t_3 a necessary condition of A's occurring at t_2 is lacking. (See the swimming pool example.) Does this mean that A is impossible relative to t_1? No. It certainly means that A does not in fact occur at t_2 (and is therefore impossible relative to t_2), but A could be perfectly possible relative to t_1 even though it did not occur at t_2. By principle (iii), of course, if a necessary condition of A is lacking at t_3, then there exists at t_3 a sufficient condition for the nonoccurrence of A, and the continuity principle (iv) stipulates that there should exist a sufficient condition for the nonoccurrence of A at every instant between t_3 and t_2. But there is no reason whatever for thinking that there should exist a state of affairs sufficient for the nonoccurrence of A at t_1, and we conclude therefore that it is quite possible for something to happen, a future necessary condition of which is lacking.

4. *Laws of Nature*

I shall conclude this part by putting forward an hypothesis which is speculative, but which I think may have some merit. The hypothesis is, that the universality of the laws of nature has as its basis physical necessity defined in either of the ways discussed in this paper.

At first sight, the situation might seem to be exactly the opposite. If I drop my pen, the necessity of its hitting the table-top might appear to be due entirely to the universality of the law of gravitation—without such a law, it might be said, no such necessity would exist. But which comes first, the necessity or the law? That is to say,

---

[6] Conversely, something can *become* necessary (assuming it is not impossible); but nothing can *become* possible.

is it necessary that *A* should occur, relative to *B*, because of a universal law of nature linking features of *A* with features of *B*, or is the law of nature a product of the physical necessity of *A* relative to *B*, together perhaps with that of *C* relative to *D*, etc.? Remember that the necessity in each case is *individual*, while the law is *general*, and that it is orthodox to regard the individual as metaphysically prior to the general. Furthermore, from what does the law of gravitation derive its universality? Day after day the moon circles the earth, providing the disciple of Newton with confirmation that bodies attract one another proportionally to the product of their masses with the inverse square of the distance between them. But why this boring continuous confirmation? Is there any reason, any necessity, why one day bodies should not attract one another according to an inverse cube law, so that the moon goes hurtling off into space? Well, if the moon's continuing to circle the earth is *necessary* relative to its present mass, position and velocity, then it *cannot* hurtle off into space. Or, if you wish, bodies not only *do* attract each other proportionally to the product of their masses with the inverse square of the distance between them; they *must*.

Admittedly the reasoning put forward here is very sketchy, and it would take a lot more sophisticated argumentation to show that laws of nature owe their universality to a sort of physical or 'real' necessity which is very different from logical necessity.[7] But such is the hypothesis that I wish to put forward for consideration.

II

5. *The Leibnizian Definitions*

It is not difficult to construct a formal system embodying the Leibnizian definitions of the physical modalities; in fact much of

---

[7] This conclusion would be entirely unacceptable to such philosophers of science as Ernest Nagel: see *The Structure of Science* (New York, 1961) , pp. 52 ff. But not perhaps to all philosophers of science. Thus 'Those who believe that empirical science can be adequately expressed in a language having the structure of *Principia Mathematica,* i.e. an object-language devoid of such modal expressions as "necessarily" and "possibly", . . . face a trying test of their faith.' Arthur Pap, *An Introduction to the Philosophy of Science* (2nd printing, New York, 1962) , p. 292. It should be clear by now that if this paper contains any implicit doctrine of causation, it will not be a doctrine based on the Humean theory of constant conjunction.

the work has already been done, notably by Arthur Prior.[8] One lets $p, q, r, \ldots$ range over propositions like 'Dion is walking' which can have different truth-values at different times, and x, y, z, range over instantaneous world-state ('slices' of the four-dimensional world perpendicular to the time-axis). We then add the primitive functions Lxy ('world-state x is later than world-state y') and Txp ('p is true in world-state x') to first-order functional calculus with identity, and add axioms for L which depict the world as an infinite tree branching toward the future but not toward the past.[9] Each branch of this tree represents a possible world-history. Relative to any point on the tree there is a manifold of possible futures, represented by the branches which ramify and proliferate above that point, and a unique past, namely the limb which leads to the tree's root. We now define 'p is possible relative to z' as 'p is true on some branch which passes through z' i.e. $(\exists x)(Bxz \cdot Txp)$, where Bxy ('x and y are on the same branch') is defined as $Lxy \lor Lyx \lor x = y$. 'p is necessary relative to z' will then be $(x)[Bxz \supset (\exists y)(Bxy \cdot Typ)]$, namely '$p$ is true on every branch which passes through z'. These two Leibnizian definitions run parallel to the 'weak' and the 'strong' definitions of futurity given in the abstract cited in the last footnote. But no more will be said about them here. Instead we shall proceed to devise formal analogues for necessary and sufficient conditions, concerning which very little work has been done.

6. *A Calculus of Necessary and Sufficient Conditions*

The primitive notions in terms of which the concepts of necessary and sufficient condition will be formalized are slightly different from those employed in the previous section. The variables x, y, z, ..., instead of ranging over instantaneous world-states, will range over event-types,[10] while the variables $t, t_1, t_2, \ldots$ range over times. In addition we shall use the primitive functions O and $>$, reading Oxt as 'an event of type x occurs at time t', and $t_1 > t_2$ as 't_1 is later than t_2'. For example, x might be the event-type 'swimming', in

---

8 *Past, Present and Future* (Oxford, 1967).

9 For appropriate axioms see the author's abstract "On What it Means to be Future," *The Journal of Symbolic Logic*, 33 (1968), p. 640.

10 Event-types are what von Wright calls 'generic events'. See his *Norm and Action* (London, 1963), p. 26.

which case Oxt would denote that swimming takes place at time t. Or x might be the event-type 'Nuel's swimming in Green Bay', so that Oxt denotes that Nuel swims in Green Bay at time t, and $(\exists t_1)$ $(t_1 > t \cdot Oxt_1)$ that Nuel swims in Green Bay at some time later that t. The primitive notation also includes the symbol $\rightarrow$ of connexive implication,[11] restricted in such a way that no 'nesting' of $\rightarrow$ is permitted. That is, in no well-formed formula $\alpha\rightarrow\beta$ do either α or β contain an arrow. The full list of primitives and rules of formation of what I shall call the system NS is as follows:

Primitive symbols
Event-variables: $x, y, z, \ldots$
Time-variables: $t, t_1, t_2, \ldots$
    Operators: $\cdot, \sim, \rightarrow, \exists, O, >$
Rules of formation
   (i) If α is an event-variable and β is a time-variable then $O\alpha\beta$ is well-formed.
  (ii) If α and β are time-variables then $\alpha > \beta$ is well-formed.
 (iii) If α and β are well-formed then $\sim\alpha$ and $\alpha\cdot\beta$ are well-formed.
  (iv) If α and β are well-formed and contain no occurrences of '$\rightarrow$', then $\alpha\rightarrow\beta$ is well-formed.
   (v) If α is an event-variable or a time-variable, and if β is well-formed, then $(\exists\alpha)\beta$ is well-formed.
  (vi) Nothing else is well-formed.

The definitions and the axioms of the system NS require some thought, and will be considered one by one. First the regular definitions of '$\vee$', '$\supset$', '$\equiv$' and the universal quantifier are made in terms of '$\cdot$', '$\sim$' and '$\exists$'. Then come the all-important definitions of necessary and sufficient conditions:

Df. 1. $Sxt_1yt_2 = Oxt_1\rightarrow Oyt_2$
Df. 2. $Nxt_1yt_2 = Oyt_2\rightarrow Oxt_1$.

Sxt_1yt_2 and Nxt_1yt_2 are read, respectively, 'The occurrence of an event of type x at time t_1 is a sufficient condition for the occurrence of an event of type y at time t_2', and (for short) 'x at t_1 is a necessary condition of y at t_2'. It follows directly from the definitions that if x is a necessary condition of y, y is a sufficient condition of x, and vice versa. We now need axioms to establish that the

---

11 S. McCall, "Connexive Implication," *The Journal of Symbolic Logic*, 31 (1966), pp. 415-433, and 'Connexive Implication and the Syllogism,' *Mind*, 76 (1967), pp. 346-356.

relations N and S are reflexive and transitive; that for example x at time t is a necessary (and a sufficient) condition of itself, (i.e. $Oxt \to Oxt$), and that if x is a sufficient condition of y, and y of z, then x is a sufficient condition of z. These properties of N and S follow from the corresponding properties of '$\to$'. Thus, where p,q,r are well-formed formulae which contain no arrow, we need

Ax. 1 $p \to p$

Ax. 2 $[(p \to q) \cdot (q \to r)] \supset (p \to r)$

Furthermore, if the occurrence of an event x is a sufficient condition for the occurrence of an event y, then the nonoccurrence of y is a sufficient condition for the nonoccurrence of x: $(Oxt_1 \to Oyt_2) \supset (\sim Oyt_2 \to \sim Oxt_1)$. And of course $Oxt \rightleftarrows \sim \sim Oxt$. These laws follow from the corresponding laws of connexive implication:

Ax. 3 $(p \to \sim q) \supset (q \to \sim p)$

Ax. 4 $\sim \sim p \to p$.

The definition of the negation of an event-type is now needed. Thus if x is the event-type 'Alan eating his soup', then $\bar{x}$ is the event-type 'Alan not eating his soup'. Since the variables x,y,z appear only in an O-context, '$\bar{x}$' may be defined contextually as follows:

Df. 3. $O\bar{x}t = \sim Oxt$.

We now come to the important thesis that if x is a sufficient condition of y, x is not a sufficient condition of $\bar{y}$. That is, nothing can be a sufficient condition of both the occurrence of an event and the nonoccurrence of the same event. In symbols, $Sxt_1yt_2 \supset \sim Sxt_1\bar{y}t_2$, which is derivable from

Ax. 5 $(p \to q) \supset \sim (p \to \sim q)$.

Axiom 5 is one of characteristic theses which distinguishes con-nexive implication from other species of implication, and represents the principal reason why '$\to$' was chosen as one of the primitives of NS. That instances of axiom 5 exactly fit the intuitive notion of what has been called 'causal implication' has been mentioned quite frequently: as early as 1947 Nelson Goodman was calling attention to the incompatibility of pairs of propositions such as the following:

If that piece of butter had been heated to 150° F., it would have melted

If that piece of butter had been heated to 150° F., it would not have melted.[12]

---

[12] Goodman, "The Problem of Counterfactual Conditionals," *The Journal of*

As a consequence of axiom 5 we may deduce that if x at t_1 is a necessary condition of y at t_2, x at t_1 is not a necessary condition of $\bar{y}$ at t_2. Manipulations of this sort require that we have at our disposal the full apparatus of two-valued propositional logic for the operators $\supset$, $\sim$, $\cdot$, $\vee$ and $\equiv$, and this we add as follows:

> Ax. 6   Any set of axiom-schemata for two-valued
> propositional logic.[13]

So far nothing has been said about rules of inference, but the following four present themselves naturally: (i) Detachment for '$\supset$', (ii) Detachment for '$\rightarrow$', (iii) Substitution of event-variables for event-variables, (iv) Substitution of time-variables for time-variables.

Next come the definitions of the physical modalities of necessity and possibility. The intuitive idea behind these definitions is that an event x at t_1 is necessary relative to the time t_2 if and only if some event y occurs at t_2 which is sufficient for the occurrence of x at t_1. Formally:

Df. 4. $Lxt_1t_2 = (\exists y)(Oyt_2 \cdot Syt_2xt_1)$,

where Lxt_1t_2 is read 'x at t_1 is necessary relative to t_2'. For physical possibility, we say that x at t_1 is possible relative to t_2 if and only if there exists at t_2 no event y whose occurrence is sufficient for the nonoccurrence of x at t_1:

Df. 5. $Mxt_1t_2 = \sim (\exists y)(Oyt_2 \cdot Syt_2\bar{x}t_1)$.

It is also not difficult to show that $Lxt_1t_2 \supset Oxt_1$ and that $Lxt_1t_2 \equiv \sim M\bar{x}t_1t_2$. For this we require

> Ax. 7   Any set of axiom-schemata and rules for
> classical quantification theory.

It is easy to show, using quantification theory, that $Oxt_1 \supset Mxt_1t_2$. The proof of the latter, for which we need an extra axiom for $\rightarrow$:

> Ax. 8 $[p \cdot (p \rightarrow q)] \supset q$

proceeds as follows:

1.	$[Oyt_2 \cdot (Oyt_2 \rightarrow O\bar{x}t_1)] \supset \sim Oxt_1$	Ax. 8, Df. 3
2.	$Oxt_1 \supset \sim (Oyt_2 \cdot Syt_2\bar{x}t_1)$	1, Df. 1, PC

Philosophy, 44 (1947), pp. 113-128. See also A. W. Burks and I. M. Copi, "Lewis Carroll's Barber Shop Paradox," *Mind*, 59 (1950), pp. 219-222, and R. B. Angell, "A Propositional Logic with Subjunctive Conditionals," *The Journal of Symbolic Logic*, 27 (1962), pp. 327-343.

13 Note incidentally that axiom 1 is derivable from 2-6.

3. $Oxt_1 \supset \sim (\exists y)(Oyt_2 \cdot Syt_2 \bar{x}t_1)$ 2, QT

4. $Oxt_1 \supset Mxt_1t_2.$ 3, Df. 5

We shall not in the sequel list any more axioms in the style of axioms 1-6 or 8 which concern propositional logic alone. Instead we shall feel free to make use in the system NS of any well-formed formula which is satisfied by the following truth matrices:

$\rightarrow$	1 2	$\sim$
*1	1 2	2
2	2 1	1

$\cdot$	1 2
1	1 2
2	2 2

These matrices provide a proof of consistency of the quantifier-free fragment of NS, together with those formulae which, like (x) $Oxt \supset Oxt$, are satisfied by the matrices when their quantifiers are stricken from them. This class of formulae, however, whose consistency is so easily demonstrated, by no means exhausts the list of desirable theses for NS. We have for example the principle of the continuity of sufficient conditions, discussed in Section 3 above:

Ax. 9 $[Lxt_1t_2 \cdot (t_1 > t_3 > t_2 \lor t_2 > t_3 > t_1)] \supset Lxt_1t_3,$

and the principle which enables us to argue from the absence of a necessary condition for the occurrence of an event to the presence of a sufficient condition for the nonoccurrence of that event and conversely:

Ax. 10. $(\exists x)(Nxt_1yt_2 \cdot \sim Oxt_1) \equiv (\exists x)(Sxt_1\bar{y}t_2 \cdot Oxt_1).$

Furthermore, many of the quantifier-free axioms that were set down earlier may be strengthened in a natural and plausible way by the addition of quantifiers. We have, for example, as a strengthening of axiom 5:

Ax. 11 $(\exists x)Sxt_1yt_2 \supset \sim (\exists x)Sxt_1\bar{y}t_2.$

Axiom 11 states that if, at any time, any event is sufficient for the occurrence of a second event, then no event at that time is sufficient for the nonoccurrence of the second event. This axiom could be strengthened even further to state that no event at *any* time would be sufficient for the nonoccurrence of the second event:

Ax. 12 $(\exists x)(\exists t_1)(t_1 \neq t_2 \cdot Sxt_1yt_2) \supset \sim$
            $(\exists x)(\exists t_1) Sxt_1\bar{y}t_2.$[14]

---

[14] Note the addition of the proviso that t_1 and t_2 be distinct. Without it, the antecedent of axiom 12 would be a thesis and the consequent therefore detachable.

It would, however, be absurd to try to strengthen axiom 5 differently; for example to $Sxt_1yt_2 \supset \sim (\exists y) Sxt_1\bar{y}t_2$ or to $Sxt_1yt_2 \supset \sim (\exists t_2)$ $Sxt_1\bar{y}t_2$. If x at t_1 is sufficient for y at t_2, it would be a mistake to conclude that x at t_1 is sufficient for the nonoccurrence of *no* event at t_2.

There are two matters which will be discussed in conclusion: conjunctive or 'molecular' events and Laplacean determinism. Molecular or conjunctive events are easy to define in the same way as negative events, namely by a definition similar to definition 3:

Df. 6. $O(xy)t = Oxt \cdot Oyt,$

and we may read off from the matrices a number of plausible laws such as $S(xy)t(yx)t$. Great care is called for, however, since in dealing with molecular events it is sometimes very difficult to distinguish what is true from what is false. For example, the following formula is satisfied by the matrices:

(K) $(Sxt_1zt_2 \cdot Syt_1zt_2) \supset S(xy)t_1zt_2.$

But is it true? Suppose for example we take x to be 'putting milk in tea', y to be 'putting lemon in tea', and z to be 'improving the taste'. Then it might well seem to be the case that x is sufficient for z, and that y is sufficient for z, but that x and y together are *not* sufficient.[15] However, it should not immediately be concluded that this example invalidates principle (K). For if putting milk and lemon into tea does not improve the taste, then it would seem that putting milk in is not sufficient for improving it after all. If we follow up this line of reasoning we see in fact that (K) holds, since every state of affairs which falsifies its consequent automatically falsifies its antecedents. Of course, it is easy to see what those who reject (K) have in mind. Imagine a pure cup of tea sitting isolated from all outside influences. Call this steady set of background conditions C. Now inject into C successively the following *and nothing else:* (i) milk, (ii) lemon, (iii) milk and lemon. In cases (i) — (ii) of this experimenter's dream we get an improvement in taste, in (iii) not. But though such circumstances are easily imagined, the formalization of them is not, and the challenge of devising a precise formal analogue of the species of sufficient condition implicit in this example (which is more of a hothouse plant than the wilder and less

---

15 Those who have ever put milk and lemon together into tea are unlikely to repeat the experiment.

sophisticated strains reflected in the examples given earlier in the paper) will not be attempted.

I shall conclude with a brief discussion of the axiom of Laplacean determinism. Universal or Laplacean determinism is the doctrine that there are no contingent events, i.e. that for any event-type x at any time t_1, either x or its negation is necessary relative to all times. Hence if we are Laplaceans we shall add the following axiom to the system NS:

Ax. 13 $Lxt_1t_2 \lor L\bar{x}t_1t_2$.

Another way of reading axiom 13 is: given any event-type x and any time t_1, either there exists at all times a sufficient condition for the occurrence of x, or there exists at all times a sufficient condition for the nonoccurrence of x. It is useful in this connection to define a formal operator for *contingency*. An event-type x at t_1 is contingent relative to t_2 if and only if both x and $\bar{x}$ are possible relative to t_2:

Df. 7. $Cxt_1t_2 = Mxt_1t_2 \cdot M\bar{x}t_1t_2$.

Then the Laplacean axiom asserts that no events are contingent, i.e. $(x)\,(t_1)\,(t_2) \sim Cxt_1t_2$.

It is almost impossible to believe that the axiom of universal determinism is true. It would imply, for example, that there existed a million years ago a sufficient condition for my now writing this sentence. In fact, it destroys the very difference between necessity and actuality by allowing us to prove that every event that occurs is necessary:

1. Oxt_1 — hyp
2. $L\bar{x}t_1t_2 \supset O\bar{x}t_1$ — see above
3. $\sim L\bar{x}t_1t_2$ — 1, 2
4. $Lxt_1t_2 \lor L\bar{x}t_1t_2$ — Laplace
5. Lxt_1t_2 — 3, 4
6. $Oxt_1 \supset Lxt_1t_2$. — 1, 5

We shall, therefore, cast about for alternatives to axiom 13. These alternatives will assert that at least some events are contingent with respect to some times. There is a difficulty, however, in the fact that there exists a very large number of alternative axioms to choose from. They range from the very strong (and false) axiom that all events are contingent with respect to all times—$(x)\,(t_1)\,(t_2)\,Cxt_1t_2$—to the very weak assertion that some event is contingent with respect to some time—$(\exists x)\,(\exists t_1)\,(\exists t_2)\,Cxt_1t_2$. In fact there are 26 such axioms in all, arranged according to their prenex quantifiers in the following table:

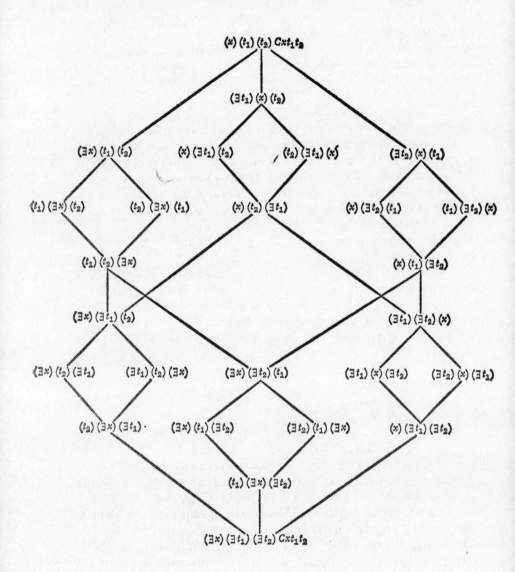

The downward-pointing lines indicate relations of implication. Which axiom or axioms should we choose? Well, if there were in the world such things as genuinely random or chance events, for which there existed no causal antecedents or sufficient conditions whatsoever, then $(\exists x)\,(t_1)\,(t_2)\, Cxt_1t_2$ would be a likely candidate. And perhaps in such things as flips of a coin, or in the disintegration of individual atoms of a radioactive substance, or in the initiation of causal chains of electrochemical events in the brain which eventually lead to the movement of a part of the body, such events are to be found. But caution is called for. Despite its attractions the assertion $(\exists x)\,(t_1)\,(t_2)\, Cxt_1t_2$ is unacceptable as it stands, because not even the flip of a coin at time t_1 is contingent with respect to *all* times t_2: it is not contingent for example with respect to t_1 itself. Relative to the present, in fact, all present events are determined (either positively or negatively), and we need a weakened form of Laplace's axiom to state this:

Ax. 14 $Lxt_1t_1 \lor L\bar{x}t_1t_1.$

We derive as a consequence of axiom 14 that $(x)\,(t) \sim Cxtt$. Secondly, it seems odd to say that any event-type should be contingent relative to the *future*. Relative to 1968, for example, is it both possible that Marlowe wrote *Hamlet* and possible that Marlowe did not write *Hamlet*? Epistemically, perhaps, but ontologically no. Hence if we are to accept $(\exists x)\,(t_1)\,(t_2)\, Cxt_1t_2$ as an axiom it must be subject to restriction, and what I propose is:

Ax. 15 $(\exists x)\,(t_1)\,(t_2)\,(t_1 > t_2 \supset Cxt_1t_2).$

Axiom 15 implies a corresponding thesis in the system NS for each expression in the lower left-hand half of the table given above. These theses become very weak, however, as we progress toward the bottom of the table, and for existentially quantified t_1 and t_2 we need the stronger versions 15a. $(\exists x)\,(t_2)\,(\exists t_1)\,(t_1 > t_2 \cdot Cxt_1t_2)$ and 15b. $(\exists x)\,(t_1)\,(\exists t_2)\,(t_1 > t_2 \cdot Cxt_1t_2)$, which follow from axiom 15 with the help of

Ax. 16 $(t_1)\,(\exists t_2)\, t_1 > t_2$
Ax. 17 $(t_2)\,(\exists t_1)\, t_1 > t_2.$

Axiom 15 states that some event-types are contingent at all times with respect to all earlier times; 15a states that some event-types are contingent with respect to all earlier times at some time or other; and 15b states some event-types are contingent at all times with respect to some earlier time. This completes our discussion of

Laplacean determinism and the alternatives to it, and with it our discussion of the physical modalities.

STORRS MCCALL

UNIVERSITY OF PITTSBURGH
AND
MAKERERE UNIVERSITY COLLEGE
KAMPALA, UGANDA

TIME REPRESENTED AS SPACE

The practical arts demand that the men who practice them have well-defined objectives. Any set of practical objectives in turn requires a maximum inattention to other considerations. A man hitting a tennis ball will fail, if his attention is distracted by a passing funeral procession, or the flutter of a handkerchief in the crowd watching him. The thief is likely to thwart his own ends, if he thinks too concretely about the consequences of his act.

It is the perennial task of philosophy to reopen doors that we have closed in our headlong pursuit of our practical needs. We can not avoid, nor should we, our preoccupations with what is practical. But there are many modes of practice. Each kind of practice, be it playing tennis or the saying of a Requiem, is a filter that excludes part of the world for the sake of ordering and controlling the rest. There is no mode of practice whose presuppositions are adequate for generating a total philosophy. From the demands of practical business life, for example, no adequate ethics can be derived. From a devotion to religious literature—no matter if that be one of the most practical things we can pursue—no suitable account of the material world can be derived. From the practical love in a family no model for political action can be constructed, even if the hippie generation thinks so. And from the powerful conception of justice in political society there can be no image made which will satisfy the family needs of the developing child, even if Plato thought so, as he apparently did.

Each of these phases of our existence, personal, social, material, religious, and commercial, is an abstraction from all the rest. The great danger in these restricted enterprises is success. Success in one's own particular practice convinces him that he has got his hands on the primary reality. And therefore the more he will argue that other visions of reality are best tested by one's own particular discipline. The rest of this essay is addressed to the vanity which holds that the abstract considerations of material science provide an adequate framework for understanding our experience of temporality. The particular target chosen is the spatial representation of time.

I. Suppose I draw a line, moving my chalk from left to right and finishing the drawing with an arrow point showing the "direction" of time. What have I shown? The gesture is autobiographical for my culture. I have shown primarily a habit of mind which men have fallen into and by which I understand something. The diagram has purely human significance. Cats have been betrayed by images, bees by colors, and plants by artificial light. But only in relation to human beings does this diagram signify. The diagram is objective in the sense that it is not arbitrary and private. It has meaning for others and could have its range of meaningfulness extended even farther.

The diagram does not merely represent the basic intuition of time as extensive, but as passing, moving, etc. The character of extension time shares with space. But the character of movement is peculiar to time alone. The diagram may exhibit this continuing character by a string of dots to the right, in the fashion of one indicating an indefinitely continuing mathematical series. "More to come," these dots say. If I draw lines to make a triangle, it does not matter what the order of drawing the lines is. I can do as I please with spatial presentation. But for the time-line, so called, directional order, though arbitrarily chosen, once decided can not be altered *ad lib*. I may prefer to have the time-line running from bottom to top, but then I must cling to that convention. This is why I put an arrow point on the diagram, not merely to give a *direction* to time, but to exhibit the fact that the time-line, although simply standing there now, was laid out by an act embodying extensive *motion*. Normally I can grasp the drawn line in a single glance. Psychologically, therefore, I am permitted to neglect this primary sense of temporal movement which underlies all of our other conceptions of movement. It might be pedagogically sounder if I never drew a time-line short enough to be seen in a glance. Then, since I should forever have to be using a discernible length of time in which to pass over the length of the time-line, I might remember that any effort to represent time simply as space neglects the primary awareness of time as *passing*. One type of abstraction in the representation of time as space is quite clear; passage of time is only given by suggestion and is easily forgotten.

A second kind of abstraction given in the spatial representation of time is evident in the omission of the additive character of time. The description of time as "moving" or "passing" does not of itself warn us of the fact that time seems to be involved in growth without loss. Insofar as time is both extensive and real, it seems to be adding to itself. Last Monday's events were not actual; now they are actual and past. The "additive" character of time is derivative from its "moving" character. One might be wiser to say that the "movement" and the "addition" both derive from the basic conception of "passage." However we conceive of space—finite, infinite, or neither—we do not find in it this factor of autonomous growth. Time as its own source of increase contains many paradoxes and challenges. It is somewhat scientifically offensive, since it apparently introduces the notion of activity without any agent, and this involves the idea of *causa sui*, long unpopular in the material sciences. However, Newton was wise enough to build into the "mathematical" notion of time this quality of self-production. Such time, he says, "flows equably and from its own nature."

A third kind of abstraction given in the diagram of time as a time-line, is what I shall call its "isolation." By 'isolation' I mean that time is dimensionally divorced from the spatial dimensions, by the very act of being represented in one (or two) of them. When a thing is "represented," the genuine article is not being given. Love shown as a red-painted heart thereby is represented by something distinct from it. In actuality there is little, if any, sense of the concrete passage of time that is not drawn from raw materials that are at least vaguely spatial as well. Such isolation is hardly treacherous in itself, of course. Much the same isolation is accomplished when I deal with the apple's taste as distinct from its shape. In my diagrammed line I often "restore" the spatial dimensions somewhat, by drawing on one-dimensional line at right angles to the time-line, or by representing a plane at right angles to it. Thus one dimension is made to stand for three, with or without the mediating suggestions of two dimensions.

There is, finally, a fourth kind of abstraction in the spatial representation of time. This abstract time is itself no prime element in our experience. Neither the empty space nor the empty time of Newton has direct reference to any self-distinguishing element in our experience. What is given in experience is the continuum of events shared by consciousness and the world. Only by neglecting the qualitative aspects of these events are we able to abstract the conception of pure space and time. Time and space are not found in the interstices of qualities, nor vice versa. The two are given in the unity of events. To separate the one from the other is analogous to a chemical distillation, rather than a mechanical separation. Accordingly, I shall call this kind of abstraction that of distillation, suggesting the notion of a purified and rarefied essence not found in untampered-with nature.

There are, then, four distinguishable ways in which our encounter with temporality must be submitted to abstraction before we can present it as a spatial line. Other—here less important—abstractions can be ignored. The fact that the line on the blackboard is not a line but a dominantly linear solid, for instance, has some significance for the tactics of representation in general, but does not directly bear on the subject of time given spatially.

II. The foregoing analysis is not an indictment; it is merely a dissection of what in point of fact occurs. If what was said in the first few paragraphs of this short study is true, some practical gain is given in each of the modes of abstraction.

(1) Why the first abstraction; why deprive time of its passage? What is to be gained? The answer lies in the nature of analysis. For us to analyze normally means to have an object for analysis. No object, nothing to analyze. Every analytic act presupposes an objectifying act. Only a narrow segment of the temporal passage falls into perceptive purview. The rest is recalled and symbolically reinstated so that the present perception is not disjunctive. Passing time as represented by a nonpassing line is thus itself only an extension and rarefaction of the acts by which the perceptive act is kept from being meaningless. Perceptual meaning arises only through the joint retention of temporally distinct elements of our experience. The aim of retention is to conserve the raw materials of meaning from erasure by the passage of time. The practicality of this automatic retention extends into every field of human thought, thus claiming a heavy and ineradicable usefulness. The presentation of time as a line is a sophisticated symbol of an action which evidently is present in much less complex animals than men. The action underlies all data of associationist psychology, for example. Of any two elements of our experience that are "associated," *both* must be

objectified to be associated. The associating act transcends the items associated and requires the objectification and retention of each of them. When abstract time itself becomes the object of attention the static medium of space is the obvious theatre for its symbolic representation.

(2) Why do we abstract from the additive character of time? Just as the abstraction itself grows out of its predecessor, so does its practical usefulness. The objectification which analysis requires must yield not only a nonlapsing object but a finite one. The additive character of temporal passage requires us to think of time as open-ended. The simple cautionary note is that we never represent time itself at all, but some stretch of time, some segment, some part. But the terms 'stretch', 'segment', and 'part' all seem to be borrowed from the vocabulary of space itself. Thus we spatialize in order to finitize. It should be noticed in passing that the claim that we have represented "time" is valid only if there is a presumption of homogeneity of the segment with the (incomplete) whole of time itself. Thus the first two modes of abstraction in the spatial representation of time constantly confront one another. The elimination of the passage of time leads to a representation of time in terms of a segment of it, and the finitizing of the segment of time leads us back to the elimination of the passage. Yet these two modes of abstraction are not identical, nor do they even entail one another. For example, the motion of the second hand on a clock (even a clock face with no metric marks) is spatial insofar as the motion is watched against a backdrop of a relatively unchanging quality in space. But we feel closer to the real passage of time because the object observed is not terminating. We thus have a spatialized presentation with some of the authenticity restored by—on principle—unending motion. Plato's definition of time as the moving image of eternity is itself a nonspatial representation aimed at the same target. Although the definition *invokes* something of a visual image, likely, it is not *itself* such an image as that given by the clock face. The present participle 'moving' restores our sense of open-ended continuity.

(3) The *isolation* of time also has its practical justification rooted in perception. Perception objectifies and retains—for the primitive act of association—elements of our flowing experiences in an abiding space. Our knowledge arises from the reliability with

which we seize the recurrent features of our experience. Space is the theater in which the drama of endurance is acted out. We never encounter pure passage of time in our nascent experience. But the reflective acts in which we identify qualities and associate them results in the emergence of stabilized objects in a fluctuating field of perception, yields a by-product. The by-product is time, passing, adding to itself, and nonspatial. We live in an experiential field of passing events, permeated through and through by a remorseless temporal movement. From this ceaseless flow we differentiate objects which even if they themselves change, do so at a "slower rate." Our perception plucks enduring objects from this flow—changing or not—and enables us to differentiate ourselves from this process of change. The subtraction of objects and selves from the temporal passage brings us asymptotically closer to pure temporal sequence, the unstoppable series of moments which erodes the most persistent object, the most heroic self-assertion. Time as isolated is admittedly represented in a spatial medium in the diagram. But the remaining parts of the diagram, if it be a spatio-temporal diagram, are used to show the radical distinction between space and time. The isolation of time is the aftermath and result of the isolation of objects and subjects.

(4) Finally there is the distillation of time. Just as there is a certain intimacy between the first two modes of abstraction, so there is between the second two. The distillation of time is one step further along the road pointed out by the isolation of time. When we take away sensory content and spatial extent from the given in experience, we are left with the emotive and evaluative aspects of that experience and its passage. If we will further drop out the emotive and evaluative aspects, we are on the threshold of obtaining the pure distillate, namely pure passage. The concept of pure passage is, however, a bit ephemeral. Our chalk line on the board compromises the ephemerality of pure time in the interest of not losing it entirely.

Indeed, it is the business of the chalk line to restore some small portion of what has been lost in each mode of abstraction. Thereby is the concept of time restored to intellectual credibility. The passage is restored faintly by the act itself. For instance, a wise classroom instructor will draw his line for the student rather than having it prepared for him when he enters. The assumption of homogeneity (albeit what exactly is homogeneous is difficult to say, since the past time itself is not repeatable) of the segment with a larger incomplete whole can be given by trailing the line off indefinitely or by using dots. The isolation of time is remedied in some degree by the use of a spatio-temporal diagram. The assimilation of space and time to one another in relativity theory has multiplied the use of such diagrams, for example. And the very use of the long, slender, white solid, that is the chalk mark, is itself a warning that time void of qualitative content can at best be thought, not encountered.

III. Is there a collective justification for the joint use of all modes of abstraction? I believe there is; it lies in the need for measurement in the natural sciences and the derivative need to express the possibility of measurement. Spatial measurement is based on the assumption of an invariant and manipulable unit of measurement, a ruler of some sort. Temporal measurement is based on the assumption of a regularly invariant process, a clock of some sort. But the obtaining of either type of measurement requires a temporal process of counting—as Kant showed. Lest the items counted be themselves lost in the steady replacement and evaporation of the present we must have a mode of transposition, some kind of metamorphosis which will keep the seriality of the measured lapse of time essentially static. We need not employ a graphic line of course. We very frequently do not. We designate by such marks as '10^2 years', 'a century', '36524 days', and so on. But since the same tactic of counting is possible for spatial stretches and for temporal stretches, once the unit of measurement is given as repeatable and invariant, spatial representation becomes both possible and convenient. A wide variety of measurements from graphs of economic tides to statements of fairly complex algebraic expressions then emerges.

But the use of these essentially mathematical metaphors for time considered exclusively in terms of its metric is fraught with misunderstanding. Our practical needs for measurement and representation of measurability are profoundly abstract, like all single practice. Measurement is almost as hopelessly partial as an approach to reality as is the marketing of peas. The convergence point in the over-intellectualization of temporality is the conception of time as monolinear.

A monolinear time emphasizes the closed features of experience.

Each segment of the time-line thus represented is objectified, homogeneous in kind with all the rest, unalterable and inaccessible, save for purposes of the abstract portrait given in the diagram. The great question is this: *What aspects of the actuality from which it was derived does this diagram apply to?*

If I consider the line *qua* mathematical line, its parts are indefinitely numerous. I can have as many as I like. But we encounter no finite temporal stretches that correspond to this mathematical permissiveness. If on the other hand the line stands for a lapse of *perceived* time, then its segments are extended points (unlike the extended points of the mathematical line) and there is a finite number of them in any finite stretch. They will be somewhere between 10^{-1} and 10^{-2} seconds long, depending on how you evaluate experiments in perceptual sensitivity. If I attempt to fit my abstract monolinear time over other aspects of actuality which ride on the backs of perceptual units of time but are larger than they, I must deal with even larger temporal units. Indeed there seem to be as many ways of breaking up the temporal lapse as there are human projects and thus as there are human "nows." Irretrievability and unalterability are the characteristics of the past. Open-endedness and incompleteness are the characteristics of the present. The shot that killed John Kennedy is irreversible. That event in its sheer physicality is closed. Kennedy's hopes for his country come in terms of larger purpose involving larger nows occupying larger chunks of open-ended time.

Actually, as we pass, analytically, from the succession of eye blinks and wisps of sounds and smells which are the atoms of our perceptual world, through the somewhat abstract level of mere actual perception toward the concrete level of value and purpose, the indivisible temporal units become larger and larger. Even at the relatively low and quite abstract level of fact, our perception taken in milli-seconds is perception devoid of all meaning. If a perception has meaning—as we have already seen—it is in relation to other and previous perceptions somehow both preserved in their past temporal setting, but here before us now for purposes of comparison.

Let the least element of significance or meaning enter the temporal world and time conceived as monolinear has begun to be transcended. Every associative act brings the temporally remote and the just-now into a new now-present synthesis. And beyond the level of this often automatic transcendence of sheer serial monolinear time there looms the realm of value, with its deliberate and reflective synthesis of what is physically past, but axiologically present, into *one* current challenge to action. Here, consequently, the "nows" are of giant size.

Human lives and lived time cut across a number of times, mathematical, physical, perceptual, and valuational. Each of these times can be superimposed on all the rest. They can be all submitted to purely mathematical dissection into as many tiny unreplaceable and closed movements as you like. But when one asks, "What, in reality, does this monolinear line correspond to?" one finds only the abstract assumptions of the fine art of physical measurement. At this point, if we must use the spatial diagram, it would seem to be better to let in a second dimension—vertical, if the former diagram be of a horizontal line. In this vertical diagram we should have a number of time-lines, each one above the next, ascending in an order depending on the mathematical size of their least units. These lines would be metrically equivalent but atomically different. The basic and most abstract line would be the mathematical one itself, at the bottom, functioning as a metric framework, or reference base, for all the rest above it, but otherwise hardly representative of the reality from which it is drawn.

Human existence, from which all sense and meaning for temporal passage of any sort are derived, would then be shown as straddling these time-lines, not confined to any one of them. What would be discerned as completed and closed for one level could then be shown as still present and open in another. The specious conflict between the presumed fixity of physical fact (with or without Heisenberg Uncertainty is unimportant in this context) and the openness of the domain of human value thus evaporates. The presence of the human attitudes of hope, in consequence, or those virtues of courage and temperance which Plato placed so high in the realm of unaltering forms, could then be shown to be genuinely germaine to an objective human world. They would not be thought of as mere shared states of inward minds. For the higher, more concrete realm of value, temporal units come in very extended segments when they are considered against the scale of mathematical time. A particular courageous act may take minutes, days, years, and yet be but one act. It is not decomposible into a sequence of microscopically small courageous acts that share the homogeneity of courage. The courage emerges only in the sustained poise toward a valued end. The component facts that undergird the single value unit may be widely various and do not necessarily have to follow a fixed order. The courageous impulse may have to submit to a variety of factual challenges and reversals. It does not thereby lose its identity and it is not necessarily identifiable in kind with any of its temporal "parts." Or if any of these "parts" be "courageous," it will be in virtue of its status in the indivisible whole.

I conclude that the monolinear, monodimensional representation of time is the fusion product of a cluster of four abstract-needs for temporal understanding embodied in the practical art of measurement. I have argued that human time is but poorly represented in this one limited aspect of human understanding—understanding through measurement. Further, that a more concrete approach to time does not confine itself to the factual, the physical, and the mathematical—these modes of human understanding that demand a metric approach—but rather extends to the realm of value, intent, and purpose as well. Finally, that we see here minimal units of time—discernible in terms of the meaning for 'now' within that stratum—which are mathematically much larger than, for example, minimal units of so-called "perceptual time" which underlie them. Hence if we are to confine ourselves to the somewhat risky representation of time as space, we at least ought to do so two-dimensionally, thus giving ourselves room for an array of time-lines according to systems of "nows," and coming closer to human experience than we can with a notion of time built upon the restrictive needs of the act of measuring alone.

NATHANIEL LAWRENCE

WILLIAMS COLLEGE

ON POSTULATES FOR TEMPORAL ORDER

In Prior's [4], Appendix A §4 and §5, and Chapter IV, and more explicitly in Bull's [2], we find sequences of tense-logical systems which place increasingly more restrictive conditions on the temporal relation ". . . is before . . ." (or, equally, on its converse ". . . is after . . ."). We give here a simple, diagrammatic account of the way in which the successive postulates for temporal order place these conditions on the temporal relation: we do not provide any essentially new semantics for these systems, nor do we prove any rigorous metatheorems. We do, however, quite often assume the completeness results of Lemmon, Bull and others without explicit mention.

All of the systems to be discussed contain the operators and tautologies of classical propositional calculus, the propositional calculus rules of *modus ponens* and uniform substitution, together with the two primitive tense-logical operators G and H. These are to be read as 'It will always be the case that' and 'It always has been the case that', respectively; their duals F and P, to be read as 'It will (sometime) be the case that' and 'It has (sometime) been the case that', are defined by

$$Fa =_{df} \sim G \sim a \text{ and } Pa =_{df} \sim H \sim a \quad ;$$

and we also define L, here to be read as an operator asserting omni-temporality, by

$$La =_{df} Ha \& a \& Ga \quad .$$

All of the systems contain the rule of omnitemporalization (in the future)

RG. $\vdash a \Rightarrow \vdash Ga$

and the mirror-image rule

$$\text{MI.} \vdash a \Rightarrow \vdash S_G^X \quad S_H^G \quad S_X^H \quad a |||$$

where X is any symbol not in a,

and we have the following axioms for possible addition to the basis already set up:

1. $G(p \supset q) . \supset . Gp \supset Gq$
2. $PGp . \supset . p$
3. $Gp . \supset . GGp$
4. $G(p \supset (Gp \supset q)) .v. G(Gq \supset p)$
5. $Gp . \supset . Fp$
6r. $GGp . \supset . Gp$
7. $L(Gp \supset PGp) . \supset . Gp \supset Hp$
6i. $L(Gp \supset p) . \supset . Gp \supset Hp$

These axioms are due to a variety of authors: they are basically those of Bull [2], renumbered, and with simplifications due mainly to Prior. Notice that by RG, adjunction and the definition of L, we have the derived rule: $\vdash a \Rightarrow \vdash La$, that is, all of our tense-logical principles are themselves omnitemporal. But there is no contention within these systems that they are logically necessary in any wider sense than this. The minimal tense-logic K_t, due to Lemmon, has just axioms 1 and 2: this is the system (except that it has half as many axioms and rule MI) of [4, §4.1]. The further systems are:

Cocchiarella's system CR (relativistic causal system) ([4, §1.3])
$= K_t + $ Axiom 3,

Cocchiarella's system CL for linear time ([4, §5.4])
$= CR + $ Axiom 4,

Scott's system CS for linear, nonending and nonbeginning time ([4, §5.5]) $= CL + $ Axiom 5,

Prior's system GHl for linear rational time ([4, §5.6])
$= CS + $ Axiom 6r,

Cocchiarella and Bull's system GHlr for linear real time
$= GHl + $ Axiom 7,

Bull's system GHli for linear integral time
$= CS + $ Axiom 6i.

The ascriptions of ownership to the various systems are not meant to indicate absolute Priority of axiomatization or completeness proofs: rather they are meant to indicate in a general way the origin and articulation of the systems in the literature. All of the systems except GHli lie in a strictly ascending sequence: GHli lies on a branch from CS, as the numbering of the axioms indicates.

We demonstrate the effect of these axioms by using the method of assigning values, suitably extended from classical propositional calculus so that tense-logical principles can be treated. To illustrate

one of our conventions, consider the following working for a formula
of propositional calculus:

$$
\begin{array}{ccccccccccc}
p & \supset & q & :\supset: & q & \supset & r & .\supset. & p & \supset & r \\
1 & 1 & 1 & 0 & 1 & 1 & 1 & 0 & 1 & 0 & 0 \\
7 & 1 & 8 & 0 & 9 & 3 & 10 & 2 & 5 & 4 & 6
\end{array}
$$

In the first row, the 1's and 0's are truth-functional assignments; in
the second row, the numbers indicate the logical order in which these
assignments are made. In what follows '#n' will abbreviate 'assignment number n', and we will also write '$a = 1$' (or '$a = 0$') to
abbreviate 'wff a is assigned truth-value 1 (or 0)'. #0 is our original
assignment, representing our assumption that the formula to be
tested is false (in some row of the truth-table). All the succeeding
assignments follow from the truth-table for $\supset$, and finally #6 and
#10 are contradictory, showing that our original assumption is false
and that the formula is a logical truth.

In order to extend the method of assigning values to tense-logical
formulae, we consider not just one set of assignments, as for propositional calculus, but a system of sets of assignments, each one representing the truth-values assigned to the formulae involved at a
particular instant or time-slot. We will denote such a system of sets
of assignments by T, and individual sets of assignments within it
by t^i, where $1 \leq i$. The assignments within each t^i must obey the
truth-tabular conditions of classical propositional calculus. There
will also be a relation B on the set T : B represents the relation
". . . is before . . .", and the properties of B represent temporal
ordering properties. Now our prescriptions for assigning values to
wffs of the form Ga and Ha are:

(a) Iff $Ga = 1$ in t^i, then $a = 1$ in all t^j in T such that $t^i B t^j$,
and

(b) Iff $Ha = 1$ in t^i, then $a = 1$ in all t^j in T such that $t^j B t^i$.
The only difference between the two prescriptions is that between
'$t^i B t^j$' and '$t^j B t^i$': this is a semantical reflection of the mirror-image
rule MI. From these prescriptions, together with the definitions of
F, P and L, informal quantificational equivalences and the fact that
all t^i's must obey classical truth-tabular conditions, we may derive
the following:

(c) Iff $Ga = 0$ in t^i, then there is a t^j in T such that
$t^i B t^j$ and $a = 0$ in t^j,

(d) Iff $Ha = 0$ in t', then there is a t^j in T such that
     t^{j}Bt' and $a = 0$ in t^j,

(e) Iff $Fa = 1$ in t', then there is a t^j in T such that
     t'Bt^j and $a = 1$ in t^j,

(f) Iff $Pa = 1$ in t', then there is a t^j in T such that
     t^{j}Bt' and $a = 1$ in t^j,

(g) Iff $Fa = 0$ in t', then $a = 0$ in all t^j in T such that
     t'Bt^j,

(h) Iff $Pa = 0$ in t', then $a = 0$ in all t^j in T such that
     t^{j}Bt',

(i) Iff $La = 1$ in t', then $a = 1$ in t' and in all t^j in T
     such that either t'Bt^j or t^{j}Bt',

and

(j) Iff $La = 0$ in t', then $a = 0$ either in t', or in some t^j
     in T such that either t'Bt^j or t^{j}Bt'.

Now we may see in what sense Lemmon's K_t is a *minimal* tense-logic: it is minimal in the sense that it places no restrictions whatever on the relation B. To see this, we investigate the axioms of K_t using the method of assigning values. For Axiom 1 we proceed thus:

$$
\begin{array}{ccccccc}
\text{G} & (\text{p} & \supset & \text{q}) & .\supset. & \text{Gp} & \supset & \text{Gq} \\
1 & & & & 0 & 1 & 0 & 0 \\
1 & & & & 0 & 3 & 2 & 4 \\
\hline
 & 1 & 1 & 1 & & 1 & & 0 \\
 & 8 & 6 & 9 & & 7 & & 5
\end{array}
$$

t^1

t^2

In this, #0 is our initial assignment, representing the hypothesis that Axiom 1 is false. From #0, #1 - #4 follow truth-functionally: then #4 and prescription (c) require us to set out a new set of assignments t^2, such that t^{1}Bt^2, and in which $q = 0$. We draw the arrow to indicate that t^1 has B to t^2, and in t^2 we make #5, which satisfies the requirement of #4. Now by (a), #1 and #3 require #6 and #7 in t^2, from which #9 follows truth-functionally, and #9 contradicts #5. Hence Axiom 1 is a logical truth according to the semantical prescriptions, embodied basically in (a) and (b), which we have laid down.

For Axiom 2, we proceed thus:

$$
\begin{array}{ll}
\begin{array}{l}
1 \\
3 \\
\hline
PGp \supset p \\
1\ 1\ \ 0\ \ 0 \\
1\ 4\ \ 0\ \ 2
\end{array}
& \begin{array}{l}
t^2 \\
\\
\\
t^1
\end{array}
\end{array} \quad .
$$

#1, by (f), requires t^2 and #3. We draw t^2 above t^1 so that our earlier-later sequence shall run from the top of the page to the bottom (as far as this is possible). Since t^2Bt^1, by (a) #3 requires #4, which contradicts #2. Hence Axiom 2 is a logical truth.

The rules RG and MI of K_t are verified by our prescriptions. RG, in effect, asserts that if a wff a is unfalsifiable in any given t^1 within a system T, then it is unfalsifiable in any member of T, and this is indeed the case since no members of T are singled out for special mention in our semantical prescriptions. MI is verified because of the entirely symmetrical character of the prescriptions for G (and F) with respect to those for H (and P). Hence all the theses of K_t are logical truths, without our placing any requirements at all on the properties of the relation B on the set T.

The system CR for relativistic causal time is gained from K_t by adding Axiom 3, $Gp \supset GGp$. Working for this axiom is

$$
\begin{array}{ll}
\begin{array}{llll}
Gp & \supset & G\ \ G & p \\
1 & 0 & 0 & \\
1 & 0 & 2 & \\
\hline
1 & & 0 & \\
5 & & 3 & \\
\hline
1 & & 0 & \\
6 & & 4 &
\end{array}
& \begin{array}{l}
t^1 \\
\\
t^2 \\
\\
t^3
\end{array}
\end{array} \quad ;
$$

here #2 requires t^2 and #3, by (c), and again #3 requires t^3 and #4, by (c). Now #1 requires #5 in t^2, by (a); but unless B is transitive, so that t^1Bt^2 and t^2Bt^3 entail t^1Bt^3, #6 is not required by #1 and (a). Hence Axiom 3 is independent of K_t, since it may be falsified if we follow the prescriptions for K_t alone; conversely if we require that B should be transitive, then #6 is required and contradicts #4, so that under these conditions Axiom 3 is a logical truth. In short, Axiom 3 is the axiom for transitivity of the relation ". . . is before . . .".

CR is a "relativistic" system by default only: it includes no axioms which have a specifically relativistic flavour. For instance, it certainly contains no axiom which states the Special Principle of Relativity—this is far beyond the expressive power of the system. Its most significant omission is any axiom requiring that the relation B should be (weakly) connected: in special relativity, the relation ". . . is causally (or absolutely) before . . ." is not weakly connected, since according to the Special Theory of Relativity it is possible for two events to be such that it is physically impossible for them to be causally related (see e.g. [1, Fig. 27-2 and Ch. XXVIII]), in which case one is neither causally before nor causally after the other. In [4, p. 203] Prior argues, in effect, that CR is in fact slightly too weak for the causal time of Special Relativity; this is because the forward light-cones of any two events must *eventually* intersect, so that for any two events there must be at least one event which is in the absolute (or causal) future of both. The axiom which requires this is $FGp \supset GFp$; its operation is given by

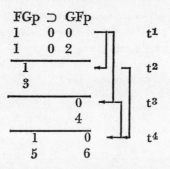

in which t^4 is the "time-slot" or event which is in the future of both t^2 and t^3. Without this t^4, #3 and #4 would not give rise to contradiction, but in t^4, #5 and #6 are contradictory. Thus the axiom is a tense-logical truth if this "ultimate convergence in the future" requirement, which seems appropriate for Special Relativity, is made.

Note that even though this "relativistic causal" relation is not connected, it is still the case that, given the temporal orderings under the Special Theory of Relativity simultaneity conventions, for a particular unaccelerated frame of reference, ". . . is before . . ." is a connected relation for events located with respect to such a frame of reference.

Thus we now turn to an examination of Axiom 4, which yields Cocchiarella's system CL for linear time. This axiom will not be a logical truth unless B is connected (over T), as we see by considering the following working:

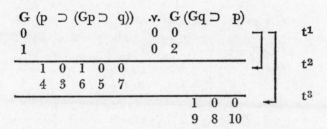

$G (p \supset (Gp \supset q))$ .v. $G (Gq \supset p)$

#3 in t^2 and #8 in t^3 are required by #1 and #2 respectively, by (c). Because of #4 and #10, t^2 may not be identical to t^3. If B is not connected, and we have neither t^2Bt^3 nor t^3Bt^2, then we have falsified Axiom 4 (and have shown it to be independent of CR). But if B is connected, then we must have either t^2Bt^3 or t^3Bt^2. If t^2t^3, then #6 and #10 are contradictory, by (a), and if t^3Bt^2 then #9 and #7 are contradictory, again by (a). That is, if B is required to be connected, then Axiom 4 becomes a logical truth, and so this axiom is an axiom for (weak) connectivity of the relation ". . . is before . . .".

The next system in the sequence is Scott's system CS, which contains Axiom 5, for the nonending and nonbeginning of time. Its operation is shown by:

$$Gp \supset Fp$$
$$\begin{array}{ccc} 1 & 0 & 0 \\ 1 & 0 & 2 \\ \hline 1 & & 0 \\ 3 & & 4 \end{array}$$

t^1

t^2

t^1, and #0, #1 and #2, constitute a falsifying instance for Axiom 5, in the absence of any condition on B. B is not required to be reflexive, so if $\sim(t^1Bt^1)$, and $T = \{t^1\}$, then #1 and #2 satisfy (a) and (g) by default. But if we add the requirement that B should satisfy the nonemptiness condition

$$(t) (t_\epsilon T . \supset . (\exists t') (t'_\epsilon T \& tBt')),$$

then Axiom 5 becomes a logical truth. For, under this condition, t^2 is required, and then #1 and #2, by (a) and (g), require #3 and #4, and these are contradictory. So Axiom 5 is an axiom for the nonemptiness condition on the relation B.

Axiom 5 is not, however, an axiom of infinity, since the nonemptiness condition is not sufficient to ensure that the set T is infinite. If B (or any of its powers) is reflexive, it will satisfy the nonemptiness condition, whether or not the set T is infinite. Hence the reflexivity of B must be excluded, at least, in order for Axiom 5 to be an axiom of infinity, and this is not done in any of the systems we are considering. In particular, both GHlr and GHli, and hence all the systems we are considering, will bear the "Smiley interpretation" in which the tense-operators G and H are interpreted as the identity operator. Prior calls this "instantaneous time," and the fact that none of the systems rule it out is a further proof that Axiom 5, although it is an axiom for nonending time, is not an axiom for infinite time.

The system GHl is now formed by the addition of Axiom 6r to CS. This axiom is the converse of Axiom 3, the axiom for transitivity. Now B is transitive if and only if $B^2 \subseteq B$: the converse of this condition is $B \subseteq B^2$, or, in quantifier notation

$$(x) (y) (xBy \supset (\exists z) (xBz \& zBy)).$$

Provided that the quantifiers in this condition are taken as ranging just over the set T, it asserts that the ordering of T by B is a dense ordering, and so we see that Axiom 6r is a natural axiom for the density of the temporal order. Diagrammatically, we have

$$
\begin{array}{ccc}
GGp & \supset & Gp \\
1 & 0 & 0 \\
1 & 0 & 2 \\
\hline
1 & & 0 \\
4 & & 3 \\
\hline
1 & & \\
5 & &
\end{array}
\quad
\begin{array}{l}
t^1 \\
\\
t^2 \\
\\
t^3
\end{array}
$$

here t^3 is required in order to satisfy the nonemptiness condition for t^2 (and we may assume t^3Bt^3 to satisfy this condition for t^3 itself). Now if the set $T = \{t^1, t^2, t^3\}$ were not required to be densely ordered by B, then we would have falsified Axiom 6r by the above construction. t^2 would be the next moment after t^1, and in it p is false, so Gp is false in t^1. But in t^3 (and all succeeding time-slots), p is true, so Gp is true in t^2 and GGp is true in t^1, falsifying Axiom 6r (and so proving its independence from CS). However, if density is required, then $(\exists t)\,(t \epsilon T\ .\&.\ t^1Bt\ \&\ tBt^2)$. This t cannot be t^1, for then #1 and #2 would be contradictory; nor may it be t^2, for then #4 and #3 would be contradictory. Hence this t is strictly between t^1 and t^2: in it we must have Gp $= 1$, by #1 and (a), and this assignment then contradicts #3 in t^2. Hence, if T is densely ordered by B, Axiom 6r is a logical truth.

In the preceding paragraph we have referred to "the ordering of T by B," and strictly speaking this is incorrect since B is not necessarily an ordering relation. To remedy this situation, we may speak of the ordering of T by B $\cap$ J, where J is the relation of diversity, if we need to use the properties of ordering relations as these are normally defined.

Our next axiom is Axiom 7, an axiom for "Dedekindian continuity." An ordered set is continuous if it contains no gaps or jumps. If it is dense, then it contains no jumps, so since GHlr is an extension of GHl we need not consider the possibility of the set T containing jumps in its ordering by B (or B $\cap$ J). But we do need to consider the possibility of gaps. These are defined in [5, XI.5]: a gap in T is a pair of disjoint nonempty subsets, say a and β, of T, such that $a \cup \beta = T$ and such that one of the subsets, say a, is such that all of its elements bear relation B to any element of β, and such that the subset a has no last element under the relation B and the subset β has no first element under the relation B. We want to show that Axiom 7 may be falsified if T has a gap, but that if T has no gaps then Axiom 7 is a logical truth. Our working to show this proceeds thus (we choose the mirror-image of Axiom 7 just to make the diagram easier to draw):

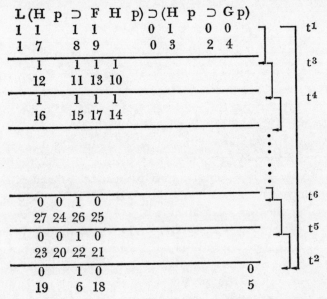

Here #4 requires t^2 and #5, and then #1 and (i) require #6. (The reason for drawing t^2 at the bottom of the diagram will be apparent when we consider the other t''s). Then #8 is required by #1 and (i), and #9 is required truth-functionally. In turn, #9 requires #10 and t^3. Now we may not have t^2Bt^3, for then #5 and #10 would be contradictory; hence by connectivity of B we must have t^3Bt^2. In t^3, #11 is required by #1 and (i), #12 repeats #10 since this is the assignment $Hp = 1$ in t^3, and thence #13 follows truth-functionally, and thence t^4 and #14 by #13. t^3 and t^4 may be seen to be the beginnings of an endless (not necessarily infinite) series, each member t^j of which is such that t^jBt^2. But there is also an endless series leading backward from t^2. In t^2, we must have #18, which sets $FHp = 0$, for otherwise if $FHp = 1$ in t^2, then this will require $Hp = 1$ in some t^x where t^2Bt^x, and this will contradict #5. Another way of seeing the justification for #18 is to observe that $FHp \supset p$ is the mirror-image of Axiom 2, and we already have $p = 0$ in t^2. Given #18, #19 follows from #6: this in turn causes t^5 and #20, and this causes #21 in the same way as #5 caused #18. This begins this endless series, each member t^h of which is such that t^jBt^h, where

t^j is any member of the previous endless series. Hence the two endless series form a gap in T: if T can have such a gap, then Axiom 7 may be falsified, but otherwise Axiom 7 is a logical truth. Hence Axiom 7 is an axiom for Dedekindian continuity (or, rather, for a weakened form of Dedekindian continuity appropriate to sets which need not be infinite).

Finally, we proceed up the other branch of our set of tense-logical systems, to the system GHli for linear integral time. 'Integral' in this context means that the set T under the ordering B is ordinally similar to the natural numbers or some (finite nonnull) subset of them. This is equivalent to the requirement that every cut in T should be a jump; that is that for every pair of disjoint nonnull subsets of T, say a and β, such that one of the subsets, say a, is such that all of its elements bear relation B to any element of β, the subset a has a last element under the relation B and the subset β has a first element under the relation B. We now show that if there were such a cut in T for which a had no last element, then Axiom 6i could be falsified; and if β had no first element then the mirror-image of Axiom 6i could be falsified. For the first case, we have the construction

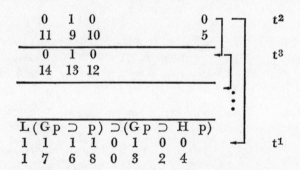

An endless series is begun from t^2: the mechanism whereby #10 and #9 cause #11, and then #12 and #13 cause #14 creates this endless series in the forward direction. No member of this series may be t^1, because of #8 in t^1. Hence if $a = T - \{t^1\}$ could be an endless series, then Axiom 6i could be falsified, but otherwise not. Similarly for its mirror-image, we have

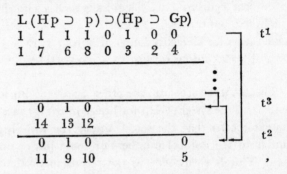

and thus the mirror-image of Axiom 6i, and via rule MI Axiom 6i itself, may be falsified if $\beta = T - \{t^1\}$ has no first member. Hence Axiom 6i is an axiom for integral time.

For a careful reading of the manuscript of this paper and consequent improvements therein, I am obliged to Ian Hinckfuss.

M. K. RENNIE

UNIVERSITY OF QUEENSLAND

BIBLIOGRAPHY

[1] D. Bohm, *The Special Theory of Relativity* (New York: W. A. Benjamin, 1965).

[2] R. A. Bull, "An Algebraic Study of Tense Logics with Linear Time," *Journal of Symbolic Logic*, 33, No. 1 (March 1968), pp. 27-38.

[3] A. Church, *Introduction to Mathematical Logic*, Vol. 1 (New Jersey: Princeton University Press, Princeton 1956).

[4] A. N. Prior, *Past Present and Future* (Oxford: Oxford University Press, 1967).

[5] W. Sierpiński, *Cardinal and Ordinal Numbers* 2d ed. (Warsaw: PWN, 1965).

HERE AND NOW

One of the most puzzling things about time is that peculiar experience we all have of the present forever "moving" from the past towards the future. What is now future becomes progressively closer to the present as time goes on, until it becomes present, and finally slips away into the past. Philosophers of time seem to divide themselves into two main camps concerning the ontological status of these phenomena. The objectivist insists that this temporal "becoming" is an objective feature of the real world, that this progression of *now* is an aspect of reality quite independent of our experience. The subjectivist argues to the contrary that temporal becoming is a subjective phenomenon which has no existence apart from the experience of some sentient being. Richard Gale in Chapters X and XI of *The Language of Time*[1] takes the objectivist position. He argues for the objectivity of temporal becoming by claiming that the conceptual systems embodied in ordinary language rule out the subjective position.

I feel that although much of what Professor Gale says on the subject is illuminating, there is a giant obstacle for an attempt to demonstrate the objectivity of temporal becoming along the lines that he proposes: an obstacle which he recognizes, but one which I do not think he successfully surmounts. Gale puts his finger on the problem in the following passage:

> This objection, which I am sure the reader has wanted to raise at several places in this book, assumes that there is no objective here in nature. I grant this assumption. This objection then attempts a *reductio ad absurdum* of my claim by showing that all of the considerations I have advanced for the objectivity of A-determinations [determinations with respect to past present and future] also hold true for here.[2]

It is clear that if we isolate some truth about *now* and hope to

---

[1] (London: Routledge & Kegan Paul, 1968) . Hereinafter cited as LT.
[2] LT, p. 213.

offer it as evidence for the objectivity of temporal becoming, we must be sure that the same feature is not also true of *here* lest we commit ourselves also to the objectivity of spatial becoming, which I assume no one would want to do. Thus the establishment of the objectivist thesis depends to a great extent on the interesting problem of the similarities and dissimilarities between the concepts *here* and *now*. The challenge posed to Gale by our obvious commitment to the subjectivity of *here* is to demonstrate that *here* and *now* are dissimilar, and dissimilar in ways which bear on the proof of his thesis. Any remark about *now* which is also true of *here* must be irrelevant to the demonstration of the objectivity of temporal becoming.

Although Gale recognizes this challenge in word, I do not think he does so in deed, since he devotes so little time to showing the differences between *now* and *here*, but concentrates instead on showing the purported failure of the Russelian and Reichenbachian analyses of tense. According to these analyses 'X is now ϕ', is equivalent to 'The ϕing of X is simultaneous with this mental experience/utterance'. Since the analysis of *now* makes explicit reference to either a mental experience or an utterance, it follows that the concept of *now* is necessarily subject dependent.

Gale replies that the analyses fail, because the analysans and the analysandum do not have the same entailment relations. The analysans entails the existence of either an utterance or a mental experience, while the analysandum clearly does not. It can easily be the case that it is now ϕ although no utterance to that effect is made, and although no one experiences ϕ. Thus while it may be the case that in *using* a sentence of the form 'X is now ϕ', the statement made is true just in case the ϕing of X is simultaneous with that utterance, the *meaning* of the statement does not entail that there is an utterance. Gale admits that as a pragmatic matter the truth of such a sentence is dependent on the time of its utterance, and so he says that it is *pragmatically* subjective, but he insists that as far as the *meaning* of the sentence goes, there is no reference to a speaker or experiencer, and hence is semantically objective. We must not become victims of Gale's use of the word 'objective' here and conclude that because sentences containing 'now' are semantically objective that *this* can serve in any way as evidence for the objectivity of temporal becoming. Just as expressions containing 'now' depend for their truth on the context of utterance, and yet do not entail the existence of an utterance, the same is true of expressions containing 'here'. This just points out that the proof of the objectivity of temporal becoming does *not* depend on the proper theory of indexical expressions. Whatever is true, according to such a theory, of the indexical 'now' will also be true of 'here', thus rendering all results of such a theory irrelevant to demonstrating the objectivist thesis, since any support for that view must be in the form of properties which are true of *now* but false of *here*.

Gale realizes that it behooves an objectivist to show significant disanalogies between *here* and *now,* and he does attempt to do so in *The Language of Time.* However, he makes a consistent error in his analysis of the differences, an error which is easily made, but one which must be avoided if the issue of the objectivity of temporal becoming is not to be prejudiced in his favor. Demonstrations of the dissimilarities between *now* and *here* can be thought of as having the following basic form. A sentence containing 'now'—let us call it N—and a corresponding sentence containing 'here' (called H) are produced. It is then shown that while one of the two is a necessary truth, the other is merely contingent, i.e. that N and H differ in logical status. But this sort of demonstration will be spurious if careful attention is not paid to getting the proper pair of sentences for analysis. We must be sure that the two sentences produced are exact analogues of each other. It clearly will not do to argue that *here* and *now* are fundamentally different by pointing out that:

N_1 All present events happen now.

is tautologous while:

H_1 All present events happen here.

is contingent. H_1 turns out to be contingent because we have retained a temporal interpretation of the word 'present' in our reading of it. Hence H_1 and N_1 are not analogues since not all temporal concepts in N_1 have been replaced by their spatial counterparts in H_1. To remedy this we must make explicit the temporal interpretation of 'present' in N_1 and replace it with the appropriate analogue:

N'_1 All events which happen now happen now.

to which the obvious analogue is:

H'_1 All events which happen here happen here.

Note that where H_1 was contingent, H'_1 turns out to be a necessary truth exactly like N'_1. Once we locate the proper pair of sentences for comparison, the differences between *now* and *here* disappear.

In order to avoid such bogus disanalogies some sort of restrictions must be put on the formation of analogues. Two conditions seem fairly obvious. The first (which we have hinted at above) is that in forming the analogue of a sentence N (containing 'now'), all temporal concepts must be replaced by counterpart spatial concepts in the analogue H. We cannot allow temporal concepts in N to remain in H since the result may be found to be contingent or even meaningless simply because temporal and spatial categories are oddly mixed, and not because of any fundamental differences between space and time. Careful attention must be paid to this requirement because it is so easy to overlook tensed verbs in N. We must not forget that tensed verbs make temporal determinations and so must be traded in for appropriate spatial expressions when forming analogies. I stress this point because Gale consistently biases his discussion of the differences between *here* and *now* by failing to eliminate tense in the formation of spatial analogues. Once the proper analogues are formed according to this first principle, all the examples which Gale cites as evidence for the objectivity of temporal becoming turn out to be just as true of *here* as they are of *now*. The second condition is less central to this discussion since Gale does not violate it explicitly in forming analogues. It is that if H happens to contain temporal concepts as well as spatial ones, they must be replaced by the corresponding spatial concepts in N. This principle can be shown to be a consequence of the first principle together with the extremely intuitive requirement that the analogue of the analogue of a sentence A must be A itself.[3] We may guarantee that all of the conditions discussed are met by requiring simply that two analogues must be such that spatial concepts appear

_____

[3] This can be shown as follows: let us denote a sentence containing 'here', and spatial and temporal concepts by H (st). Suppose we *fail* to follow the second requirement and take $N(t't)$ as our analogue of H (st), (where $N(t't)$ consists of the sentence containing 'now' where H contained here, temporal concepts t' corresponding to the spatial concepts s in H, and the same temporal concepts as in H). Then the analogue of N $(t't)$ will have to be H (ss') if we accept the first condition (where s' denotes the spatial concepts corresponding to those denoted by t in N). Thus the analogue of the analogue of H (st) turns out not to be H (st), but rather H (ss').

in the first exactly where the corresponding temporal concepts appear in the second, and vice versa.

Having armed ourselves with some simple principles of analogue construction, we may proceed now to survey Gale's remarks about the differences between *here* and *now* which he feels point to the objectivity of temporal becoming. Gale commits a classic violation of the first principle in the following passage:

> This difference between here and present or now is due to the fact that there is no spatial analogue to temporal becoming: the present (now), unlike here, shifts inexorably, independently of what we do. Every event later than the present will become present and every event earlier than the present did become present, to which the spatial analogue would be that every object in front (to the right, etc.) of me will occupy (become) here and every object in the rear (to the left, etc.) of me did occupy here. But whereas the former is necessarily true, the latter is contingent, and what is more is almost certainly false.[4]

What Gale overlooks is that:

H_2 Every object in front of me will occupy here and every object in the rear of me did occupy here.

is not the proper analogue of:

N_2 Every event later than the present will become present and every event earlier than the present did become present.

because we have not replaced the temporal concepts *will be*, and *did* with their spatial counterparts in H_2. Gale has smuggled in temporal notions in the formulation of H_2, and it is no wonder that H_2 turns out to be contingent. If we reformulate N_2 so that 'did' and 'will be' are replaced with their definitions in terms of earlier than now and later than now respectively, we may easily construct the correct spatial analogue. Thus N_2 becomes:

N'_2 Every event later than now *is* present at a time later than now and every event earlier than now *is* present at a time earlier than now.

(Where the italicized 'is' represents a tenseless sense of 'is'.) To which the analogue is:

H'_2 Every event to the front of here *is* here (or spatially present)

4 LT, pp. 214-15.

at a place to the front of here and every event to the rear
of here *is* here at a place to the rear of here.

This formulation mirrors exactly the tautologous nature of N'_2,
and thus there is no disanalogy between *now* and *here*—on this
account.

Gale makes the same mistake in further remarks intended to
show that 'now' represents an objective feature of experience independent of our actions. He points out[5] that:

N_3 My next utterance of 'now' will necessarily denote a different time (from my last utterance of 'now').

is a necessary truth, while the purported analogue:

H_3 My next utterance of 'here' will necessarily denote a different place (from my last utterance of 'here').

is contingent, since I may choose to stay put if I like. This time he
has included temporal notions in H_3 in the form of 'next' (and
'last'). We must reformulate N_3 to make these temporal references
clear:

N'_3 My future utterance of 'now' which is closest in time to
the present necessarily denotes a different time from that
denoted by my past utterance of 'now' which is closest in
time to the present.

The analogue for this will be:

H'_3 My utterance of 'here' to the front which is closest in position to here necessarily denotes a different place from that
denoted by my utterance of 'here' to the rear which is
closest in position to here.

Here again we note that when the proper spatial analogue is formed,
the statements concerning *now* and *here* have the same logical status.

Temporal concepts get smuggled into the spatial analogue in a
more subtle way in the following passage:

Also our use of tensed or A-questions—questions that require
an A-statement [a statement containing tense determinations]
for their answer—presupposes that the communicants share the
same temporal but not the same spatial perspective. This is
brought out by the fact that it is meaningful to ask "Where
are you?," to which the answer could be "I am here."; herein
it is the direction that the voice comes from that constitutes the

---

5 LT, p. 214.

answer. However, there is no use for the question "When are you?": "I am now" has no use in our language.[6]

The trouble is that:

N_4 It is meaningful to ask "When are you?" (and to answer "Now.")

is not the proper analogue of:

H_4 It is meaningful to ask "Where are you?" (and to answer "Here.")

The reason is that tense distinctions are packed into the verb 'are' in H_4. It is certain that 'are' here cannot be read tenselessly. When we ask "Where are you?," we do not want to find out what position the person questioned *did* or *will* take, we want to know where he is *now*. Thus we must rewrite H_4 as:

H'_4 It is meaningful to ask "Where *are* you now?" (and answer "Here.")

and the correct analogue of H'_4 is:

N'_4 It is meaningful to ask "When *are* you here?" (and answer "Now.")

But it is just as meaningful to ask "Where *are* you now?" as it is to ask "When *are* you here?" (to which the answer might be "Every-day at nine."), hence there is no difference in the logical status of H'_4 and N'_4.

A final consideration which Gale points to in trying to establish his thesis ties in with the general theme that although we have control over where *here* is going to be, we seem to have none whatsoever over where *now* is going to be. Gale expresses himself by saying that while:

H_5 ". . . I can, within certain physical limits, choose the place denoted by my use of 'here' n time-units from now."[7],

it is not the case that:

N_5 ". . . I can, within certain physical limits, choose the time denoted by 'now' n space-units from here."[8]

But there are difficulties in evaluating the truth of N_5. At first

6 LT, p. 215.

7 LT, p. 214.

8 *Ibid.*

glance it seems to be simply meaningless—it makes no sense at all to choose a time denoted by 'now'. Yet we can, I think, make sense of 'choosing the time denoted by "now" *n space-units from here*'. I *can* make it be the case that at a certain place *n* units from here that a use I make of 'now' denotes a preselected time. All I must do is get to that place in time and utter 'now' just at the proper instant. So there is a way of making sense of N_5 so that it comes out to be just as true as H_5 was. If we recast H_5 a little, then we will be able to construct a meaningful temporal analogue:

> H'_5 I can make it be the case (within certain physical limits) that my use of 'here' *n* time-units from now denotes a preselected place.

the analogue to which is:

> N'_5 I can make it be the case (within certain physical limits) that my use of 'now' *n* space-units from here denotes a preselected time.

Clearly there is nothing impossible about doing this. I have just as much control over what time my use of 'now' denotes at a certain place as I do over what place my use of 'here' denotes at a certain time. The view that I have control over where here is and none over when now is seems plausible because the temporal analogue is formed incorrectly. We ought to express matters more carefully by saying that I have control over where the place denoted by 'here' is *going to be*. The analogue to this is not: I can control when the time denoted by 'now' is *going to be*. This is simply meaningless, and it violates the principle that in forming analogues, expressions like 'going to be' must be replaced by their spatial counterparts. It is just because we do not pay particular attention to tensed expressions like 'will be', 'was' and 'going to be' that we seem forced to say such strange and paradoxical things about time.

So it appears that Gale has failed to meet the challenge posed by the similarities in the concepts *here* and *now*.[9] While I have not

---

9 Gale does offer another argument on p. 215 which must be handled a bit differently. There he says that testing "for the objectivity of our present sense experience, which is based on the agreement in judgment between different observers, presupposes that these observers exist at present and therefore share the same temporal perspective; but they need not, and usually do not, share the same spatial perspective, that is they are not all standing at the same place." But is it really the case that the only observers relevant to the assessment of the objectivity

demonstrated the *subjectivity* of temporal becoming, I hope to have shown the kind of difficulty which objectivists like Gale must meet if they are to argue successfully *for* their position. Gale and others of his persuasion may try to defend themselves against the arguments of a subjectivist, and this is done with a good deal of cogency in *The Language of Time*. But if the objectivist is to try to bring positive evidence for his thesis, he must invariably clear the hurdle posed by the similarities between space and time. So far, I have not seen any purported evidence for the objectivity of *now* whose proper analogue is not also true of *here*. Until evidence is found which does not have this deficiency, the objectivist will necessarily find himself in a very weak position.[10]

JAMES W. GARSON

UNIVERSITY OF PITTSBURGH

---

of my present experience share the same temporal perspective? Surely an observer's past determination that there were no elephants in the area yesterday is relevant to the question of whether my present experience of an elephant in my living room is objective or not. Far from being a conceptual truth, the view that observers testing the objectivity of present experience must exist now is false, as is its *proper* spatial analogue: that observers testing the objectivity of the experience of what is here must exist here.

[10] I wish to thank Richard Gale for the interest he has shown in this paper, and for the helpful criticisms and constant encouragement he has given me.

THE FALLACY OF CONJUNCTIVE ANALYSIS

My purpose in this paper is to examine a pitfall in empiricistic analysis which has not been widely discussed, perhaps because it lies implicit in what may seem a harmless facet of such analysis. The kind of analysis I have in mind is analysis of any variety which seeks to reduce understanding of any object, concept, event, institution, or whatever, and its appropriate properties, to understanding of discrete elements and their properties out of which the analytically reduced can be constructed. Let us call the entity analysed into elements the *molar object,* and restrict *elements* hereafter to some set of discrete elements of some molar object which have been obtained by an analysis of the kind to be discussed. What I will have to say will apply most directly to the reductionist analyses of an older logical empiricism, but I hope to make it clear that more recent linguistic analyses encounter the same difficulty.

In order to sharpen the discussion I want to focus on the following line of thought: If a molar object has been analysed into a set of elements such that it can be shown that the molar object consists of nothing but the elements given in the analysis, then anything which may be said about the molar object can be said equivalently solely in terms of the elements and their properties. A statement this bald will not usually be found as an assumption in a given analysis, yet it seems that many analyses can be shown to make such an assumption. I intend to argue that this assumption is, in general, false; and I suggest that any analysis based on it in an unrestricted fashion commits an error which may conveniently be called *the fallacy of conjunctive analysis.* The road to the fallacy is broad and easy, for it is sufficient to analyse a molar object into a set of exhaustive elements, and then take the attitude that every property of the molar object must somehow be reducible to properties of the elements. This is just next to committing the fallacy, for it is easy to show that such construction is not always possible, and in fact is frequently demonstrably impossible when it is further assumed (as it seems often to have been) that the result of analysis will yield a truth-functional statement composed solely of statements about certain elements and their properties as the equivalent of every statement correctly attributing some property to the molar object. Briefly, what is often missing is something I will call *integration of the elements,* which in its employment may require much more powerful techniques than truth-functional constructions.

In saying this, it may seem that I am laying the groundwork for re-admitting empirical analysis so long as more powerful mathematical techniques are permitted as analytical tools. This is not my purpose, since I would deny, but I will not argue this here, that these more powerful techniques are analytic in any sound philosophical sense. What I am driving at is the following. The issue of the fallacy of conjunctive analysis is obviously closely tied to discussions of reducibility, unity of science, and such traditional topics as whether or not the whole is greater than the sum of its parts. With respect to the last, I am arguing that insofar as statements about the whole cannot be reduced to conjunctions (or truth-functions) of statements about the parts of the whole, the whole *is* greater than the sum of its parts. Philosophers have argued this against empiricistic analysis before, and I think correctly. But these arguments have often turned on features of particular empiricistic analyses, such as whether or not there can be *discrete* elements which are capable of existence independently of the particular molar object under analysis. What I want to do in this paper is to show that even if it is granted that a molar object can be analysed into a set of discrete elements, a pitfall lurks for the unsuspecting analyst in putting the elements back together. In order to show that this line of thought is not directed solely at molar objects whose elements are straws, I will indicate before the conclusion of the paper that any of a wide variety of specific epistemological analyses which dot the literature seem to have committed the fallacy that I am discussing.

To this point, nothing has been said about the ontological status of either the molar object or its elements in an analysis. For convenience, only the status of the elements will be considered, and this status only with respect to a theoretical: nontheoretical dichotomy. It will first be shown that if theoretical elements are given by the analysis, then the formal possibility of the fallacy is transparent. We will use *theoretical* roughly in the following sense: an element is theoretical at the time of the anaysis if the possibility of observing it cannot be argued for in terms of scientific or phenomenological knowledge. Since many empiricistic analyses have yielded non-theoretical elements, they will also be discussed. In this latter case it will be argued that the analyses of a molar object in many interesting cases cannot be given solely in terms of elements of the molar object, making the analysis impossible but nonetheless paving the road to commission of the fallacy if the attitude is taken that the analysis in terms of elements *must* be possible.

To discuss theoretical elements, let us consider our molar object to be a spinner, of the kind used occassionally in games. The essentials of such a spinner are a circular dial and a pointer mounted in the center of the dial which may be spun with the flick of a finger. Consider the dial to be a line defining the circumference of a circle, and suppose the bearing of the pointer to be so perfect that the resting place of the pointer after a spin cannot be predicted. Suppose further that one of the points of the dial is labelled *0*, the point opposite from it on a diameter of the circle ½, and the other points labelled as the points of the half-closed interval [*0, 1*] which has *0* as a member but not *1*. The labelling will be considered spread evenly over the circumference, e.g., the point labelled ¼ is half way between the points labelled *0 and ½* . The elements of the spinner we are interested in are the points of the dial and the point which defines the tip of the pointer of the spinner. We suppose that after a spin, the tip point is coincident with *one* (and only one) of the points of the dial. Obviously, no actual spinner is such that we could label all of the points or determine by observation which point the tip is coincident with at any moment. In this sense the elements of the analysis are theoretical, but theoretical analysis of this kind is a common procedure in science. Now the spinner has a property which we will call randomness of outcome: If the tip is placed coincident with the point labelled *0*, and spun, it is as likely to stop on any point as on any other. More formally, the probability of its stopping on some point after a spin is *1*, and for any two points *a* and *b*, the probability of its stopping on *a* is the same as the probability of its stopping on *b*. Since there are an infinite number of points, all such probabilities are indentically *0*. For if the probability of the spinner stopping on *a* were to be $\delta > 0$, then the sum of the probabliities of the spinner stopping on each point would be infinite, contradicting the assignment of *1* to the probability that the tip stops on some point.[1] The probability distribution of the tip's stopping cannot be constructed truth-functionally from the probability assignment of 0 to the tip's stopping on each point.

To see this, we may look at a similar spinner with a defective bearing. In this second spinner, unwanted friction causes the pointer to be more likely to stop in one half of the dial than in the other, but the defect is still slight enough to permit the pointer to spin freely. In consequence, the tip may still stop on any point of the dial, even though it is much more likely to stop in some intervals of the dial than in others. The molar object in this case will have a nonrandom property describing the possible resting places of the tip. Even so, the probability assignment of 0 will be made to the stopping of the tip on any given point after a flick of the finger, just as in the case of the random spinner. This is not as easy to show informally. The probability that the tip will stop on some point is 1, as in the previous case. Now suppose that the defect in the bearing were to cause the points in some interval each to have a higher probability assignment, $\delta > 0$, than the points of some other interval. Since there are an infinite number of points in any interval all of whose points are assigned probability δ', the probability of the pointer stopping in that interval would be infinite, contradicting again the assignment of 1 to the probability that the tip will stop on some point.

Let us consider the interval on both spinners between dial points 0 and $½$. If x is a random variable, we may say that the probability of an observed value of x for the first spinner lying in this interval is $½$, and suppose we say that the bearing defect is such that the probability for the same interval on the second spinner is greater than $½$. Symbolically: $Pr\delta(0 \leq x < ½) = ½$, and $Pr\delta'(0 \leq x < ½) > ½$. Since the point analysis of probabilities is the same for both spinners, it is quite clear that these molar properties of the two spinners cannot be obtained by truth-functions of probability assignments to the points

---

[1] The facts about probability distributions which are used here can be found in various textbooks on mathematical statistics. More intense discusssion of the spinner example can be found in R. Ackermann, *Nondeductive Inference* (London, 1966), or J. McKinsey, *Introduction to the Theory of Games* (New York, 1952), pp. 151-57.

of the spinners. Thus, although the dial and the tip may be regarded as nothing but certain sets of points, the points cannot be elements of an analysis since there are molar properties of the sets which cannot be expressed as properties of the points which compose them. It is possible to define these molar properties precisely within measure theory and mathematical integration associated with it, but this requires too much technique to be attempted here. For the immediate purpose of this paper, it is only necessary to indicate that the molar properties cannot be reduced to truth-functions of point elements. In the particular case of the spinner, not even quantificational methods would suffice. The random variable used to give the molar distributions is not quantifiable, since replacement of the variable by a precise point of the dial results in nonsense. $Pr\,\delta\,(0\leq x<\frac{1}{2})$ is a formal representation of the assertion that the dial point co-incident with the pointer tip of the random spinner after a particular spin has a probability of $\frac{1}{2}$ of lying between the points 0 and $\frac{1}{2}$ of the dial. Suppose we spin the spinner and its pointer tip comes to rest on the point labelled $\frac{1}{4}$. Substitution of $\frac{1}{4}$ for x in the formula results in $Pr\delta(0\leq\frac{1}{4}<\frac{1}{2})=\frac{1}{2}$, which seems to say that the probability of $\frac{1}{4}$ lying between 0 and $\frac{1}{2}$ is $\frac{1}{2}$, which is either nonsensical or false, depending on one's metaphysics.[2]

The spinner examples show that the fallacy of conjunctive analysis is at least a formal possibility. Of course no-one familiar with probability theory would conclude from the fact that each point of a set has a probability assignment of 0 that the set has a probability assignment of 0. But if one were not careful, and if he analysed the dial into the point elements which compose it, he would be ready to take the wrong step if he assumed that the molar properties of the dial must in every case be reducible to properties of the points which compose it. Lest this seem obvious, it might be noted that at least one of Zeno's paradoxes invites unconscious acceptance of the fallacy. For if points have no length, how can a line composed solely of points have length? The conjunctive fallacy (by whatever name) is surely a potential danger for analysis.

---

2 The question of when random variables may be substituted for by their particular values has played a role in statistical polemics. See, for example, Jerzy Neyman, "Note on an Article by Sir Ronald Fisher," *Journal of the Royal Statistical Society, Series B*, **18** (1956) , pp. 288-294.

Empirical analysis may be defended against the occurrence of the fallacy by arguing that the difficulties with the spinner examples are raised by the fact that the theoretical elements given in the analysis are infinite (and more than enumerably infinite). It may be suggested that analysis of molar objects into a finite number of non-theoretical elements could not be faced with analogous difficulties. To consider such analyses, we may take as a model molar object an automatic tracking system for an anti-aircraft weapon. We may consider this system to have some kind of scanning device which provides electrical impulses as the input of the system, and we will suppose that the system emits as output electrical impulses which provide firing instructions for the weapon. Since my purpose is not to discuss the issue of mentality in machines, although I do choose this example for philosophical reasons which will appear shortly, I want to take some neutral position about the significance of the machine's operations. The molar property of the tracking system which I wish to consider with respect to analysis will be described here as its (correctly) possessing the information that a hostile plane is on a certain course, a property which leads to its emission of firing instructions. This property may seem somewhat abstract, and in fact we may suppose that the molar property to be the object of possible analysis will be the specific possession at a fixed time by the system of information that a hostile aircraft is pursuing a definite course.

To simplify, it can be assumed that the hostility of the aircraft is determined by an operator which only activates the system when an appropriate aircraft is to be shot at. Further, we will assume that the scanning device is a radar-like apparatus which provides positional *fixes* to the tracking system. The fixes, in any full analysis, would be further described as various configurations of electronic equipment, the details of which it is not necessary to investigate. The analogy to the spinner case is provided by the following claim: the information provided by each fix taken by itself is zero, and the molar property of information possession attributed to the system cannot be truth-functionally constructed from the fixes alone.

It is easy to see that the system can ignore a single fix as providing no information. This is for at least two reasons. If we are talking about an actual machine, and not the theoretical model of such a machine, then intermittently defective circuitry may provide random false fix information which the system would prudently discard as of no value. Secondly, by virtue of the fact that at least two fixes may be required to initiate any computation in the system. The system may not compute, for example, unless the aircraft is approaching, rather than receding, and determination of this rough fact may require at least two fixes. This seems a perfectly acceptable example of a case in which either of two fixes, by itself, possesses no information, although taken together the fixes may provide considerable information. In general, the probabilistic reliability of actual circuitry may result in the fact that there is no fixed analysis and no exact minimum number of fixes to which a molar information property of the system is analytically reducible. Actually, the molar property will usually be quite stable in such cases under addition or deletion of a reasonable number of fixes, so that no correct analysis of the molar property, and surely no full understanding of its occurrence, is easily provided. But in our particular case, we suppose all such difficulties neatly disposed of, and a history of the particular fixes preceding the particular molar property under analysis to be provided.

The fixes themselves and any truth-functions of them, are not equivalent to the possession by the system of the molar property which would result in the weapon's firing a shell which will strike an actual plane. The point, of course, is that the fixes must be integrated correctly in a temporal sequence. For this reason, the system will have a short memory of sorts. It will ignore a fix if it is not followed in a significant period of time by another fix, and perhaps many more in a suitable pattern. And after acquiring the molar property it will forget or erase the fixes that gave rise to it. Now if the molar property is possession of the information that a plane is approaching, as opposed to merely attaining some electronic configuration, then any analysis of the system must show that some part of it is providing a suitable clock which yields the temporal information required for the integration process. But to show that the clock is suitable, that is, provides accurate enough time, we must consider some elements in the analysis which are not elements of the tracking system. In just looking at the tracking system and its elements, we could not determine whether or not the system was operating on the right time. Whether or not the passage of physical time can receive a suitable empiricistic analysis or not, the possession of molar properties by the tracking system must require for its analysis more than the elements of the system itself if temporal integration is to be a feature of its operation. The tracking system seems sufficient indication that even analysis of a molar object into nontheoretical elements may encounter the fallacy of conjunctive analysis.

In order to make good my earlier claim that this fallacy is not simply a formal invention, I would like to conclude by suggesting that all sense-data epistemologies in their traditional form seem to commit the fallacy in a manner which may be clarified by comparison to the tracking system example. There is no room here for an exhaustive taxonomy of sense-data epistemologies, but my claim is meant to be inclusive of any epistemology which claims to analyse perceptual achievement into phenomenal or physiological elements.

Suppose we start with an individual who sees (correctly) that there is a red book on a nearby table. Given a smattering of science, it is easy to propose that this seeing must be analysed into a succession of momentary retinal images, and perhaps mental operations on these images in accord with regularities projected from past experience. A common analysis of the individual's judgment that there is a red book on the table in these circumstances is that this judgment is a deductive or inductive inference from information contained in the momentary retinal images. More sophisticated forms of this general line of thinking, motivated by the apparently obvious direction that *analysis* of the situation must take, have provided the basis for many sense-data epistemologies. The argument from illusion is suggested by this line of reasoning, since the illusory seeing can be construed as a legitimate inference from premises representing appropriate elements to a false conclusion, where the elements are the same as those which could lead by another inductive inference to the correct conclusion. Now the supposition that the information represented by the correct seeing that the red book is on the table *must be* derived from retinal images or other elements which may be given by the analysis can be seen to set the stage for commission of the fallacy.

I am not arguing that we do not have sense-data of some kind during perception, but only that perceptual success cannot be constructed out of sense-data elements by either deductive or inductive argumentation. For deductive and inductive inference, as well as mathematical integration, are all timeless, in the sense that even if the order of the premises is crucial, the inferential step to a conclusion can either be shown valid on the premises or it cannot, and the time taken to set down the premises or to make the inference is irrelevant to the validity of the inference.

But integration over time at an appropriate rate is crucial to perception. It is as obvious as the fact that a movie film of human activity which is shown too rapidly or too slowly does not retain its 'human' qualities. Or that a film of a friend may capture some characteristic mannerism of his which no succession of slides could suggest. Perhaps the most dramatic example is given by certain sequences of film which are occasionally used in psychological experiments. In one such case, a few frames showing a well known movie star are spliced into the middle of a film showing a horse race. The movie star is typically not seen when the film is shown. After the viewer is told that some film of a star has been spliced into the horse race, he will recognize the star and see the film as interrupting the race temporarily. These facts seem totally inexplicable on any sense-data basis, and they indicate very strongly that what is seen is determined very largely in many cases by the observed temporal rate at which a certain perceived change is noticed. Yet the importance of temporal sequences, and the impact of their rate on what is seen, are hardly ever discussed in traditional sense-data epistemologies.

What I am suggesting is that *analysis* of a perception into sense-data may involve steps no one of which is invalid, but the result of which may be a set of elements from which the perception cannot be reconstructed without additional information, in effect at least a screening of the sense-data in the right order and the right speed. It is in this way that sense-data epistemologies seem to involve the kind of formal fallacy which has been illustrated by the spinner and tracking system examples.

I do not wish to suggest that human perception consists of molar properties of a super tracking system. The interesting comparison between the two is the way in which information represented by a molar property may be analysed into elements none of which, by itself, possesses information. This has always been a puzzling feature of sense-data analysis, and I think this comparison shows that it has nothing to do with the distinctive character of human mental activity. Sense-data analysis simply loses the temporal integrity of the molar object. If the comparison to the tracking system has any interest, it suggests that understanding of the perceptual apparatus will be dependent on a temporal sense which will require explanation in terms of at least some elements outside the immediate molar object being analysed. And this, curiously, may be a reflection of the inadequacy of a purely behavioristic and analytic study of human activity. The potential importance of this observation seems heightened by experiences with certain drugs, which demonstrate that altered perception is at least accompanied by altered temporal integration. If the above remarks are correct, altered temporal integration may well be the cause of at least some altered perception, rather than merely a concomitant, and it is easy to see how drugs might affect the temporality of brain action. My reaction to these considerations is to suppose that if many of the empirical analyses of perception in the twentieth century are to have lasting significance, they will have to be combined with an adequate phenomenological study of time.

ROBERT ACKERMANN

UNIVERSITY OF MASSACHUSETTS

THINGS

> Let a man learn to look for the permanent
> in the mutable and fleeting. . . .
>
> —Emerson

(1) 9 is necessarily greater than 7
(2) The number of planets $= 9$
(3) The number of planets is necessarily greater than 7
(4) $(\exists x)$ (x is necessarily greater than 7).

This group of sentences is used by Quine in a well-known attack on quantified modal logic and the possibility of meaningful modalized predication.[1] Modalized predication would involve specifying some object, and asserting of it that necessarily or possibly it has some given property ϕ. If this can be done sensibly, then the modal context 'necessarily ϕx' can be quantified into, and conversely. So modalized predication and quantifying into modal contexts are bound up together. Quine thinks that neither is very sensible, and says of sentence (4)

> What is this number which, according to (4), is necessarily greater than 7? According to (1), from which (4) was inferred, it was 9, that is, the number of planets; but to suppose this would conflict with the fact that (3) is false. In a word, to be necessarily greater than 7 is not a trait of a number, but depends on the manner of referring to the number.[2]

I shall not directly discuss the possibility of modalized predication in this paper. My topic will be the possibility of *temporal* predication: the specifying of something and saying of it that at some *time* it has a given property ϕ.

1. *Apparent Referential Opacity.* Temporal predication seems to be something that we do every day without getting into trouble. However, it is threatened, prima facie, by an appearance of referential

---

[1] W. V. Quine, "Reference and Modality," in *From a Logical Point of View* (New York: Harper Torchbooks, 1963), pp. 139-59. Hereinafter cited as FLPV.
[2] *Ibid.*, p. 148.

opacity directly analogous to Quine's modal example above. Consider the following:

(5) In 1966 Richard Nixon was a Republican

(6) Richard Nixon $=$ the President of the U.S.

(7) In 1966 the President of the U.S. was a Republican

(8) $(\exists x)$ (in 1966 x was a Republican).

Here, we can threaten temporal predication by constructing a Quine-like attack on quantifying into temporal contexts, as seems to be done in (8) :

Who is this person which, according to (8) was in 1966 a Republican? According to (5), from which (8) was inferred, it was Richard Nixon, that is, the President of the U.S.; but to suppose this would conflict with the fact that (7) is false. In a word, to have been a Republican in 1966 is not a trait of a person, but depends on the manner of referring to the person.

The most natural way to react to this sort of argument seems to involve the charge that it is unfair in its treatment of (7). (7) is not obviously false, but seems ambiguous between

(7S) In 1966 $(\imath x \cdot Px)\ (Rx)$ [3]

and

(7L) $(\imath x \cdot Px)$ in 1966 (Rx).

Although (7S) is false, it does not seem to be the relevant interpretation of (7). It does not single out the entity in question, Richard Nixon, that is, the President of the U.S., and make a temporal predication about him. (7L) on the other hand, does do this. These remarks are directly analogous to those made by Arthur Smullyan in his paper discussing the analogous problems for *modalized* predication. [4]

Defending temporal predication does not end with pointing out

---

3 In the notation of PM this would be written

In 1966 $\{[(\imath x \cdot Px)] \cdot R(\imath x \cdot Px)\}$,

but such an expression, particularly the second occurrence of '$(\imath x \cdot Px)$', is needlessly long and confusing. I replace this latter occurrence with just an 'x', and view the initial '$(\imath x \cdot Px)$' as a *quantifier* serving to bind it. This device is particularly useful when it is necessary to distinguish various scopes of given definite descriptions; it also captures directly Russell's view that a definite description is a kind of quantifier. The quantifier '$(\imath x \cdot Px)$' could be viewed as an abbreviation for the combination of quantifiers and predicates ' $(\exists x) . (y) :. Px: Py \supset y = x \dots$' when this binds a subsequent otherwise free 'x'.

4 "Modality and Description," *Journal of Symbolic Logic*, 13 (1948), pp. 31-7.

this ambiguity, however. For what is brought out by this defense is that temporal predication is possible just in case it is possible to quantify into temporal contexts. Both (7L) and (8) involve quantifying into temporal contexts. But this should be expected. If I specify something and say that at a certain time *it* (!) is such-and-such, I have in effect quantified into a temporal context.

2. *Identity Through Time.* Another thing that is brought out by looking at (7L) is that as a temporal predication, it involves the identity of the thing specified through time. (7L) says of the President, that is, of Nixon, that in 1966 *he* was a Republican. It says that what is today the President was in 1966 a Republican: the same *thing*, the same value for '*x*' that today is President was in 1966 a Republican.

This value for '*x*' is, one would suppose, the man Richard M. Nixon. That is, we would intuitively accept

(9) $(\exists x) (x = \text{Nixon} \cdot \text{in } 1966 \, x = \text{Nixon})$.

On the other hand, we would probably want to reject the corresponding statement about the identity of the President of the U.S. through time:

(10) $(\exists x) (x = \text{the President} \cdot \text{in } 1966 \, x = \text{the President})$.

This is false, one supposes, because it was a different man, Lyndon Johnson, a different value of '*x*' that was President in 1966. (Notice that these points about identity through time have nothing to do directly with *continuity* through time— (9) seems true just because we suppose that there is a single thing that is Nixon now and which was Nixon in 1966.)

So far, we have noticed some things about (7L) which one might expect in any discussion of tensed predication. (7L) attempts to rescue such predication from the threat of referential opacity, and thus far, (7L) seems secure. But notice how the truth-value of (7L) vitally depends on the truth of (9) and the falsity of (10). In fact this *sort* of situation is just one instance of a Fundamental Theorem of Metaphysics, a version of Leibniz's Law, that connects identity through time to scope distinctions for definite descriptions:

$$(\exists x) [x = (\imath y \cdot Py) \cdot \text{at } t \, x = (\imath y \cdot Py)] \equiv_{P,t}$$
$$[\text{at } t \, (\imath y \cdot Py) \, (\phi y) \equiv_{\phi} (\imath y \cdot Py) \text{ at } t \, (\phi y)].$$

[Provided that at *t* E! $(\imath y \cdot Py)$. Small scope, of course, for the descriptions on the left-hand side.]

Now although the assumption that (9) is true and (10) is false looks fairly straightforward, its intuitive acceptability can be questioned by a development of the idea of certain kinds of

3. *Abstract Entities.* Many things that we ordinarily say seem to involve the supposition that (10) is actually true—that there is a thing called 'the President of the U.S.' that as such has maintained its identity as a single thing for quite some time, although changing with respect to certain properties. Consider sentences like

(11) In 1956 the President was a Republican, but in 1961 he was a Democrat,

(12) In 1961 the President was Kennedy, then became Johnson, and now is Nixon.

It looks as if in saying such things we are mentioning something, a sort of *abstract* or *institutional* President of the U.S., and saying of him that in 1956 he was a Republican, but changed party and became a Democrat in 1961; that once he was Kennedy, then became Johnson, and then Nixon. It looks like we are referring to this abstract President and making temporal predications of him.

If there were such a thing as this abstract President that (11) and (12) appear to contain references to, then (10) would be true! If the abstract President were a value for our variables, then the open sentence quantified in (10) is just the sort of thing that would be true of him. Indeed, to suppose the existence of an abstract President is practically what supposing the truth of (10) (in the face of the fact that two different *men* were President) would amount to.

The danger that this creates is that allowing the abstract President to be a value for our variables would affect what was said about (7L). Since the abstract President was a Democrat in 1966, (7L) would turn out to be false instead of true. For either (i) both Richard Nixon and the abstract President are values for '*x*', in which case there is not a unique President (see section 5), or (ii) we dump Richard Nixon from our ontology, and admit just the abstract President, who was not, however, a Republican in 1966.

Before this attitude begins to look any more bizarre than it already does, it might be appropriate to begin to work backwards again and look more closely at this so-called abstract President of the U.S. One would think there should be some good reasons for refusing his admission into a reasonable ontology.

The abstract President is the sort of thing which is often viewed as a grammatical or nominal fiction. According to this view, sentences like (11) and (12) do not *really* suppose the existence of such an object. (11) for example, just means

> (11S) In 1956 the President was a Republican, but in 1961 the President was a Democrat

rather than

> (11L) The President is such that in 1956 he was a Republican and in 1961 he was a Democrat.

On this view, the pronoun 'he' in (11) is used as what Geach calls a "pronoun of laziness"; it is short for a second occurrence of the term 'the President', used merely to save the speaker the bother of repeating that term. It is not used like a bound *variable* to repeat a *reference* to a single thing antecedently specified. (There are no such things as *variables* of laziness.)

This view might continue by giving a general characterization of abstract fictions like the abstract President, and the conceptual move from (11S) to (11L) or to (11) or to (10) that seems to involve it. This sort of move is typically sneered at by Hume, who writes

> For when we attribute identity, in an improper sense, to variable or interrupted objects, our mistake is not confined to expression, but is commonly attended with a fiction, either of something invariable and uninterrupted, or of something mysterious and inexplicable. . . .[5]

In writing this, Hume was *not* thinking of anything so exotic as our abstract President of the U.S. His criticism was actually aimed at the ordinary attribution of identity to persons and everyday things through time. Hume would view Richard Nixon much as I think we would be inclined to view the so-called abstract President —as a nominal fiction, and not a *thing* genuinely identical through time. (9) would be viewed as on a par with (10) —Hume would call them equally false and misguided.

Fictions or not, Quine gives us a description of the way such ideas might come to exist. In discussing how we make a person understand what the meaning (or reference) is of the term 'the river Caÿster', he begins by supposing that

> the general term 'river' is not yet understood, so that we cannot specify the Caÿster by pointing and saying 'This river is the Caÿster.'

---

[5] *A Treatise of Human Nature*, Bk. I, Pt. 4, Sec. vi. Compare Emerson.

Suppose also that we are deprived of other descriptive devices. What we may do then is point to *a* and two days later to *b* and say each time, 'This is the Caÿster.' The word 'this' so used must have referred not to *a* nor to *b*, but beyond to something more inclusive, identical in the two cases.[6]

Here, *a* and *b* are "river stages" of the Caÿster.

Now if we get at the idea of the river Caÿster in this way, one would suppose that we might get at the idea of the abstract President of the U.S. in the same way. We point at President Kennedy in 1961, at President Johnson in 1965, and at President Nixon in 1969, saying each time, "This is the President." "The word 'this' so used must have referred not to Kennedy, Johnson, or Nixon, but beyond to something more inclusive, identical in the three cases." This thing is the abstract President. Kennedy, Johnson, and Nixon are its stages.

Let us pause for a moment and assess the course of the discussion up to this point. We began with the observation that time contexts, like modal ones, seem to be referentially opaque. That is, substitution of identicals, of singular terms that denote the same thing, seems to affect the truth-value of tensed statements. This threatened the possibility of *de re* temporal predication. In attempting to rescue temporal predication from the threat of opacity, we were led to distinguishing the scopes of definite descriptions in the manner of Smullyan. This seemed to be a natural course, but it pointed up that temporal predication involves quantifying into temporal contexts, certain assumptions about the things that are the values for our variables of quantification, and the identity of these things through time. This led to a discussion of a certain family of abstract entities which, if admitted as values for our variables, would cause trouble with our remarks about (7), and the distinction between (7L) and (7S).

Furthermore, the discussion about identity through time and abstract entities can be pushed along Humean lines to cast suspicion not only on the legitimacy of the abstract President of the U.S., but upon seemingly more concrete things like Richard Nixon.

First, a few more observations about the term 'the President of the U.S.', as it occurs in (11). The Humean view would be that (11) is to be analyzed as (11S), and that the term 'the President'

---

6 "Identity, Ostension and Hypostasis," in FLPV, p. 67.

occurring in (11S) refers to two different things at the two different times specified.

The view is that 'the President of the U.S.' is not a Name[7] for a single thing, but a definite description, which denotes at a given time whatever falls under the description at that time. A definite description is formed out of a property-term—in this case, the property of 'being President of the U.S.', and so can change in what it applies to.

And the logical rules for manipulating definite descriptions are more restrictive than for Names. We cannot, for example, infer (11L) from (11S), and we cannot even existentially generalize from the term 'the President' in (11S) and infer

(13) $(\exists x)$ (in 1956 x was a Republican, but in 1961 x was a Democrat).

To perform such inferences would be to treat 'the President' as if it were a Name—perhaps as if it were a Name for the abstract President.

So far this is a pretty straightforward collection of remarks about the term 'the President of the U.S.'. However, it can be stretched to apply to apparent Names like 'Richard Nixon', exhibiting the way one might threaten the ontological status of Richard Nixon as a single particular thing, a concrete entity persistent through time.

4. *Concrete Entities.* First, we imagine that there is a certain *property* from which we can construct the required definite description: the property of being Richard Nixon, or "Nixonizing," for short. Then the term 'Richard Nixon' will be a disguised definite description, short for 'the unique Nixonizer', or '$(\imath x \cdot Nx)$'. As a definite description, it denotes at a given time whatever falls under it at that time.

Then we may state the Humean position in terms of this definite description: in 1966 one thing Nixonized, and today a different thing does. That is, in 1966 one thing satisfied the definite

---

[7] A Name is a singular noun that always denotes the same thing. By the Law mentioned at the end of Sec. 2, we may equivalently define a Name as a singular noun whose scope in tense contexts never matters (and so might as well always be large: a scopeless 'this'). These two equivalent definitions often appear as alternative definitions of logically proper names, with t ranging over possible worlds as well as times.

description 'Richard Nixon', or took the *form* of Richard Nixon, while today a different thing does. Consequently, (9) would turn out to be false. On such a view, the inference from (5) to (8) would be as incorrect as the inference from (11S) to (11L) or to (13), and for precisely the same reason. The inference depends on supposing that 'Richard Nixon' denotes the same thing through the period of time in question—or more accurately, that the same thing satisfies this definite description at all of the times in question, as if 'Richard Nixon' were a Name rather than a definite description, naming something which Hume might characterize as the fictitious abstract Richard Nixon.

Of course looking at a term as a definite description does not necessarily imply that it denotes different things at different times. Thus, 'the unique Nixonizer' might denote the same thing now as in 1966, just as 'the President of the U.S.' denoted, one supposes, the same thing in 1965 as it did in 1968, namely, Lyndon B. Johnson.

Metempsychosis would hold that 'the unique Nixonizer' denotes something which does persist eternally through time, although it does not always satisfy that definite description: it does not always take the form of Richard Nixon. In 64 A. D. for example, it might have Neroized or taken the form of Nero. This picture is the extreme opposite of the Humean view which has a rapid succession of distinct things Nixonizing: "There is a new Nixon every day." However, *both* views involve taking names as definite descriptions. The difference is only ontological, over whether the values of the 'x' in '$(\imath x \cdot Nx)$' are eternal souls, Humean ephemerae, or what have you.

But I digress. My purpose here was to sketch a picture, or a way of looking at the term 'Richard Nixon' and at the man Richard Nixon, analogous to the way we would most likely be inclined to look at the term 'the President of the U.S.' and at the abstract President of the U.S. This picture might commonly be filled out by specifying just what Richard Nixon, or Nixonizing, is being predicated of. What is the 'x' in '$(\imath x \cdot Nx)$'? Although one might expect this to be something like a group of body cells, or possibly a soul, it need not be anything of the kind.

In fact, we might do everything backwards from our normal intuitions and ontologically admit the abstract President and reject

Richard Nixon as our persistent entities. We will, however, admit properties like Nixonizing. This results in an apparent duality between matter and form: we can picture the abstract President some years ago Kennedizing, then Johnsonizing, and then Nixonizing. We picture a single thing which changes qualitatively with respect to these properties, first having one, and then another. So the abstract President would be, for a time, one value for 'x' in the definite description '$(\exists x \cdot Nx)$'. Indeed this seems to be just the metaphysical picture suggested by (12), although it does conflict with our ordinary view of things.

These attitudes may look a bit extreme with respect to 'Richard Nixon' and 'the President of the U.S.' The first looks like a name and the second like a definite description, and it does involve quite an exercise of the imagination to pretend that one could be the other. However, I think that in many ways the difference is a matter of *degree*, for we can find in-between cases.

Suppose that ten years ago I had a pair of socks and a sweater (call the sweater 'A').[8] My mother unravels the socks and winds the wool into a ball (call the wool 'B'). Then over the years various repairs are made in sweater A, and that wool B is used to replace the old worn-away areas of the sweater A. Today, sweater A is composed entirely of wool B.

Now consider the sentence

(14)   Ten years ago I wore sweater A on my feet.

This *seems* to have a familiar ambiguity which might be analyzed as a scope-ambiguity of a definite description. If we treat 'sweater A' as a definite description, i.e., '$(\exists x \cdot Ax)$', i.e., 'the unique x that is (takes the form of?) sweater A', then we can exhibit what would be the large and small scopings of the definite description in (14):

(14S)   Ten years ago $(\exists x \cdot Ax)$ (I wore x on my feet)

(14L) $(\exists x \cdot Ax)$ ten years ago (I wore x on my feet).

First, (14S) is unproblematically false, since ten years ago there was no sweater to be found on my feet. Now (14L) might look true: it says that what is now my sweater A was, ten years ago, on my feet.

However, thus counting (14L) as true involves taking B, the wool in the sweater, as a value for 'x' and rejecting the sweater A

---

[8] I take this example from David Wiggins's "On Being in the Same Place at the Same Time," *Philosophical Review*, 77 (1968), pp. 90-5.

itself as a value for that variable. The temptation to view (14L) in this way seems to come from a temptation to read '$(\exists x \cdot Ax)$' as 'the *matter* of sweater A'.

But that seems unjustified: '$(\exists x \cdot Ax)$' was intended as short for just 'the sweater A'. Now if the sweater A itself (an abstraction?) were a persistent value for 'x', if it were such that ten years ago *it* was sweater A rather than the wool B, then (14L) would be false. This seems fairly intuitive, but notice also that such an identity across time of the sweater *qua* sweater, together with the unproblematic falsity of (14S), would yield that falsity of (14L) by the Fundamental Theorem mentioned above in Sec. 2.

Here then we have an example of the same logical type as the situation involving Nixon and the abstract President. But the problems generated in this one seem much less artificial. We must distinguish at least A (the sweater), B (the wool now in the sweater), C (the form of the sweater), and D (wool). A and B are two things, since they have distinct temporal properties. A is distinct from C, although A has C. B is in A. A is made of B. A is made of D. B is distinct from D, but B is a bit, piece, sample of D. . . .

In any case, viewing 'sweater A' as a definite description fails to automatically eliminate the problems in interpreting (14). Indeed, it just sharpens our focus on a very old philosophical problem. A and B seem to be equally good candidates as values for 'x' in (14L), but admitting *both* creates difficulties. We might look at the problem this way: suppose that this troublesome garment could *talk*, and said "Ten years ago I was on Sharvy's feet." Is it the wool speaking, and truly, or the sweater speaking falsely? Finally, why not the sweater speaking truly? Or even the wool speaking falsely?

For even the identification of *matter* is not always completely clear. I might say that I have had this beard for five years, but this hair on my face for less than a week. But in a fit of materialism I might say that I have had this beard for less than a week; and in the opposite frame of mind, I might say that I have had this hair on my face for five years. There seems to be some truth in all of these, even the last two. The task of separating the strict talk from the loose is not merely automatic here.

5. *Identity, Congruence, and Ontology.* There remains this fact: given a definite description such as 'the President of the U.S.' we

cannot admit as entities *both* the abstract President and the concrete President (Nixon), unless we are willing to let that definite
description be improper, in that then there would be not *one*
President, but *two*: the abstract President and Nixon.

Although such improper definite descriptions are fairly common,
particularly where "universals" are involved ('the act I am now
performing', 'the tenth letter in this paragraph', etc.), they are
viewed with some distaste. In the present case, one suggestion for
making the descriptions proper would involve viewing the abstract
President and Nixon as being *now identical,* and in the past distinct. Similarly, one might view the sweater A and the wool B as
now identical but formerly distinct. We can draw a picture of this
view:

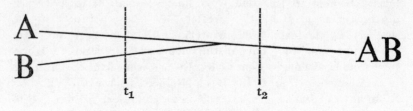

Figure 1

This is supposed to picture a pair (?) of things A and B which are
distinct at t_1, but then become identical by t_2.

I think that this is a nice picture, but I think that if the notion
of identity is to be taken seriously, we cannot allow it to be
interpreted as a picture of two things A and B being identical at t_2
and not at t_1. At best, they can be merely *congruent* in some way at
t_2.

Otherwise, time contexts would be referentially opaque, temporal predication would be impossible, and we would be back
where we started. For we could not (at present) refer to Richard
Nixon and predicate of him the property of having been in 1966 a
Republican, for something that would be identical to him, the
abstract President, lacks this property. And in general, the law

$A = B$

$\therefore$ At t, ϕ (A) iff at t, ϕ (B)

would fail unequivocally. We began this paper by looking at *ap-*

parent failures of this law—but if temporal predication is possible, the law must have a valid construal.

Another example may illustrate this point more simply. Suppose A and B in the above drawing to be *highways*, so that our drawing is a road map. t_1 and t_2 could then be lines of longitude instead of times. Highways A and B come together, and at t_2 occupy the same *road*. But can we properly say that at t_2 A and B are identical? Would it not be better to say that at t_2 A and B *congrue* or *coincide*, although they are still distinct highways?[9]

For if at t_2 A=B, then a man standing on the road at t_2 could not refer to A and say of *it* that a few miles back it is to the north of B. If at t_2 A=B, then at t_2, referring to A would be referring to B. One would be faced with the apparent referential opacity of *spatial* predicates. We could not stand at t_2 and say "The highway I'm standing on comes from the north," because something identical to it, we would have to suppose, comes from the south. Thus, in order to preserve spatial predication and Leibniz's Law, we must reject the view that at t_2, A=B.

6. *Substrata.* In an important sense, none of this is novel or problematic. In some sense it is perfectly clear what happens when highways come together, when pairs of socks are pulled apart and knitted into sweaters, and when the Presidency changes hands. What is problematic is the demand that these situations be described in a certain formal fashion. Subject-predicate language in general, and quantification theory in particular, seem to show some strain in their application to these situations.

Describing these situations in the language of quantificational logic requires that we analyze the situations ontologically, to determine what *things* are to be taken as values for the quantified variables, and what *predicates* are to be attached to them. But much of what I have said here was discussed by Aristotle, notably in the *Physics*, in *Coming-to-be and Passing-away*, and in Bks. VII and VIII of the *Metaphysics*. It is the subject-predicate view of the world that breeds the appearance of dilemma.

---

9 I borrow the term 'congruent' and perhaps some of its meaning from Quine's "The Problem of Interpreting Modal Logic," reprinted in Copi & Gould (eds.), *Contemporary Readings in Logical Theory* (New York: Macmillan, 1967), pp. 267-73. I have also discussed this and the related problems of temporary identity in "Why a Class Can't Change Its Members," *Noûs*, 2 (1968), pp. 303-14.

We could put many of our problems into Aristotelian language. Thus, the question about 'Richard Nixon' being a Name or a definite description ' $(1x \cdot Nx)$ ' might occur in Aristotle as the question: is Richard Nixon a *substratum* of which predications are made, or is Richard Nixon actually predicated of some further substratum—of the 'x' in ' $(1x \cdot Nx)$ '? (These are really all wrong ways of putting the question.) One thing we have noticed is that this further substratum need not be more *concrete* than the supposed Richard Nixon, and contrary to both Aristotle and current dogma, the substratum of change and predication for a given thing need not always be the *matter* of that thing. We considered one view in which the substratum for 'Nixon' was the abstract President, in connection with interpreting (12).

And analyzing apparently straightforward sentences like (7), (8), and (7L) immediately involves an assumption of a substratum of which predications are made. Quantification into temporal contexts, temporal predication, and identity through time: these are bound up together, and lead into what might be called the jungle of Aristotelian substratism:

> . . . there must be a *substrate* underlying all processes of becoming and changing. What can this be in the present case? It is either the body or the soul that undergoes alteration . . . [10]

That these fundamental problems arise in discussions of quantified tense logic has not been sufficiently realized. Analogies between time and modality have been explored, notably by A. N. Prior. However, Prior has concentrated mainly on time and modality in propositional logic, and less on the problems that arise with the introduction of quantification. He has not extensively examined the metaphysical implications of the apparent referential opacity of tense contexts, which invariably confront quantified tense logic. Indeed, in the last paragraph of a recent book, he states

> The problems of tensed predicate logic all arise from the fact that the things of which we make our predications, the 'values of our bound variables', include things that have not always existed and/or will not always do so. [11]

------

[10] *Physics* E2, 226a 11-3, Hardie and Gaye trans.

[11] *Past, Present and Future* (Oxford, 1967), p. 174.

This is a problem. But it actually turns out to be a species of the more general problem which I have been discussing.

One additional point of analogy between tense logic and modal logic is worth noting. Sometimes in discussions of modality the notion of a *possible world* occurs. These possible worlds have *times* as their analogues in temporal logic. One supposed problem with possible worlds is the question of identity through possible worlds. There are some very difficult questions about identity through possible worlds, often generated by considerations analogous to those which work to generate difficulties about the identity of a thing through time.[12] However, this does not impugn the notion that there *are* different possible worlds any more than the presence of thorny problems of identity through time impugns the notion that there are different times.

7. *Two Grades of Involvement.* The problems of specifying the values of bound variables and discussing identity through time cannot be avoided in an attempt to do quantified temporal logic. These problems occur because our temporal predications must obey Leibniz's Law: the Indiscernibility of Identicals. If a thing is to have a property, such as a temporalized property, it must have this property "under any description." If this cannot be justified, temporal predications must be abandoned.

Quine's feeling about the analogous problems in modal logic is that modalized predication must be abandoned. In "Three Grades of Modal Involvement"[13] he distinguishes between two sorts of modality, which we may call the *semantic* and the *predicative* (sometimes *de dicto* and *de re*). On the semantic, statements like (1) would be viewed as saying something about a statement or sentence, i.e., '9 is greater than 7'—that it is analytic. For the predicative (*de re*) grade, (1) would be viewed as saying something about a thing, the number 9, that it is necessarily or essentially greater than 7. The latter leads to what Quine calls the "jungle of Aristotelian essentialism"—the notion that some things have some of their properties essentially.

---

12 See, for example, R. M. Chisholm, "Identity through Possible Worlds: Some Questions," *Noûs*, 1 (1967), pp. 1-8, and a reply by R. L. Purtill, "About Identity through Possible Worlds," *Noûs*, 2 (1968), pp. 87-9.

13 In *Ways of Paradox* (New York: Random House, 1966), pp. 156-74.

Quine does not like Aristotelian essentialism. I am not quite certain why. Surely the apparent opacity of modal contexts could not count against Aristotelian essentialism, for only *semantic (de dicto)* modality is opaque, and semantic modality does not have anything to do with *de re* modalized predication or Aristotelian essentialism. On the other hand, predicative modality is difficult to understand in some ways, but one thing we do know about it is that it must not be opaque if it is to be predicative. And for the temporal case, there must surely be a way of understanding *de re* temporal predication.

One could of course distinguish semantic and predicative temporality. On the semantic view, sentences like (5) would be viewed as saying something about a statement or sentence, 'Richard Nixon is a Republican', namely, that in 1966 it was true. Semantic temporality is referentially opaque, because it would render (7) unambiguously false: the sentence 'The President of the U.S. is a Republican' was false in 1966.

The predicative *de re* grade of temporality has been examined. We have seen that it leads to Aristotelian substratism—the notion that there are things that persist "identically" through time having some properties at one time and other properties at other times. But if this is a jungle, it is one that we all live in.

Furthermore, it is worth noting that the dichotomy between semantic and predicative logical operators occurs in a wide variety of philosophical places. The following list of operators will generate apparently opaque contexts at noun positions of contained subsentences. These contexts will be *really* opaque if the operators are taken to be semantic, and philosophically interesting (like essentialism and substratism) if they are taken to be transparent and predicative, operating *de re:*

There is a place where P, Rasputin asserted that P, Agnew knows that P, that P is about Socrates, that P is the same statement as that Q, if P then Q, P only if Q, when P then Q, P only when Q, P because Q, it ought to be that P, I desire that P, the probability is r that P, etc.

As one more example of this sort of thing, let us look at a probabilistic case. Imagine that a certain man, Ronald, is a known petty criminal. In the past he has habitually used loaded dice that were weighted to roll a 7 half the time. (A fair pair comes up 7 one time in six.) Ronald's frequency of using loaded dice is so high that we may reasonably expect him to always be using loaded dice. Today however, he is playing with a fair pair of dice, which we may call 'D'. Now consider this triplet of sentences: 'The probability is 1:6 that the fair pair D will come up 7'; 'Ronald's pair of dice = the fair pair D'; and

(15) The probability is 1:6 that Ronald's pair of dice will come up 7.

If our probability operator is viewed as semantic, as a predicate of statements or sentences, then (15) is false, since given Ronald's habits, the sentence 'Ronald's pair of dice will come up 7' has a probability greater than 1:6.

On the other hand, if the probability operator is viewed as predicative, as forming the composite predicate *having the probability 1:6 of coming up 7* that (15) predicates of the pair of dice D, then (15) might be viewed as being true. However, this leads us into the jungle of predicating probabilistic properties *de re*. On such a view, (15) makes a probabilistic predication of the pair of dice itself. It does not make it of the pair of dice "under the description 'fair pair D'," or of the dice "*qua* being the fair pair D," or of the dice "on the hypothesis that they are fair," or in any other *de dicto* manner.

Many of the familiar problems in the philosophy of science involve this sort of intentionality of probabilistic, conditional, and causal contexts. Many of the problems raised by Goodman in these areas are caused by such factors. For example, it appears that by referring to the class of green emeralds as 'the class of grue emeralds' we can confirm a new hypothesis incompatible with the hypothesis confirmed by thinking of this as the class of green emeralds.

Now the operators listed five paragraphs ago all exhibit this sort of bifurcation between the *de dicto* and the *de re*. Their semantic or *de dicto* sides are referentially opaque. Indeed, nearly any non-truth-functional operator will go in this list.[14]

On the other hand, their *de re* sides often lead into jungles. Often, this is because the operators somehow seem to be more

---

[14] See my "Truth-Functionality and Referential Opacity," *Philosophical Studies,* 21 (1970), pp. 5-9.

semantic than predicative, and hence will appear very strongly to be opaque. But if their *de re* sides lead into jungles, at least they are traditional philosophical jungles. To prefer their *de dicto* sides in order to avoid these traditional problems seems cowardly.

I believe that much in these problems can be dealt with as one problem, that they are all species of a more general philosophical difficulty, and that the study of the logic of time yields one of the clearest models with which to approach them. But we must confront these problems, even if we must take to the jungles.[15]

RICHARD SHARVY

SWARTHMORE COLLEGE

---

[15] This paper has been stimulated by many people who have been at Wayne State University, especially Richard Cartwright, Robert Sleigh, and Lawrence Powers.

ON THE INTELLIGIBILITY OF THE
EPOCHAL THEORY OF TIME[1]

I. In "Whitehead's Theory of Becoming,"[2] V. C. Chappell asserts that Whitehead's "epochal theory of time or becoming is both untenable and unnecessary" (525; 77). It is untenable, he argues, because " (1) ... the theory itself is unintelligible, and ... (2) ... the Zenonian argument on which the theory is founded, even in its amended, Whiteheadian version, is invalid" (521; 73). Although it is not clear to me that Whitehead's use of Zeno's arguments is unsound, our concern here will be limited to the first of these two claims. I will attempt to show that Chappell's demonstration of the unintelligibility of Whitehead's theory begs the question. His purpose is to present us with a *reductio ad absurdum* of the epochal theory of time, but he succeeds only in demonstrating that if we substitute other premises for Whitehead's own, his conclusions will not be consistent with these substitute premises.

The theory is not only untenable, Chappell argues, but unnecessary. It is unnecessary because "Whitehead's basic reason for holding that actual occasions are indivisible unities, atomic drops of process, is . . . his doctrine that they are finally caused" (526; 78). Chappell attempts to divorce the "atomic character" of Whitehead's actual occasions required by his doctrine of final causes from his epochal theory of time, supported by Zenonian arguments (525-526;

[1] I am in debt to many who made time to read this paper prior to publication. Although others also left their mark upon it, Professor Donald Sherburne's comments were particularly helpful. In acknowledging his help I mean to express my gratitude to all other friendly critics as well.

This essay will be republished in the forthcoming *Festschrift* for Martin Eshleman entitled *Philosophic Essays in Honor of Martin Eshleman,* edited by Geoffrey Petersen. (Northfield, Minn.: Starr Printing, 1971) , pp. 38-50.

[2] In the Whitehead Centennial Issue of *The Journal of Philosophy,* 58, No. 19 (Sept. 14, 1961) , 516-28; reprinted in *Alfred North Whitehead: Essays on His Philosophy,* ed. George L. Kline (Englewood Cliffs, N.J.: Prentice-Hall, 1963) , pp. 70-80.

77-78) . Since one can maintain the former while denying the latter, he argues, the latter is not necessary to Whitehead's position. Whatever the status of the Zenonian arguments, however, the epochal theory of time is not, as Chappell maintains, "a minor appendage . . . which has no ground in or connection with any other element in the system" (525-526; 77-78) . Our first task will be to show that the atomic character which Chappell allows already entails the epochal theory which he attempts to discredit.

II. "The fundamental fact to which the epochal theory alludes is the indivisibility of actual occasions," Chappell asserts, "i.e., the indivisibility of the processes or acts of becoming of which actual occasions are constituted." Whitehead sometimes expresses this point by saying that the act of becoming that constitutes an actual occasion "is not extensive" (PR 107),[3] i.e., is not temporally extended. He even says that the act "is not in physical time" (PR 434) (518; 71) . Thus, "the epochal theory stands or falls with the notion of a becoming that is not extensive" (521; 73) .

It is essential that we understand what Whitehead means when he characterizes an actual occasion as nonextensive. In the passage to which Chappell refers, Whitehead explains "that in every act of becoming there is the becoming of something with temporal extension; but that *the act itself is not extensive, in the sense that it is divisible into earlier and later acts of becoming* which correspond to the extensive divisibility of what has become" (PR 107; italics mine) . Let us compare this with what Chappell identifies as one of the "major and essential tenets in Whitehead's position regarding actual occasions . . . that actual occasions are finally caused and hence have an atomic character, which is to say that *the process that constitutes an actual occasion cannot be divided into processes that are themselves actual occasions*" (526; 78; italics mine) . Since Chappell has already identified the "act of becoming" to which Whitehead refers as the process or "act of becoming that constitutes an actual occasion" (518; 71; quoted above), the two italicized statements are equivalent.

The obvious conclusion is that Chappell finds the epochal

---

3 *Process and Reality*. The pagination is identical in the Macmillan (New York, 1929) ; Social Science Book Store (New York, n.d.) , and Torchbook (New York: Harper and Brothers, 1960) editions.

theory of time one of the "major and essential tenets in Whitehead's position." As we have seen, however, he does not. Although he defines the epochal theory of time in terms of nonextensive or indivisible actual occasions, he does not use these terms as Whitehead does in the passage just quoted. In order for a theory of time to qualify as epochal, Chappell argues, it is not sufficient that occasions be composed of parts nonextensive in this sense; the epochal theory of time is "the notion that actual occasions are indivisible *because temporally nonextensive*" (526; 78; italics mine).

Whitehead has no need of *this* theory, Chappell argues, because the same actual occasion which "cannot be divided into processes that are themselves actual occasions" is nevertheless "divisible into parts that are themselves becomings of something with temporal extension" (526; 78). Chappell clearly means that the becoming of the occasion, not merely the occasion which has become, is divisible in this sense. Whitehead has no reason to deny that the occasion which has become is divisible into parts with temporal extension, nor is such divisibility incompatible with the epochal theory of time, whether defined by Whitehead or by Chappell (see Chappell: 518; 71). Furthermore, in saying that these parts are "becomings of something with temporal extension," he must mean that each part is itself something with temporal extension. He cannot mean merely that the sum of the parts, i.e., the whole actual occasion, constitutes something with temporal extension if it is to be a denial of Whitehead's position (cf. Chappell: 525; 77).

What Chappell wishes to deny is Whitehead's claim that the process by which the occasion becomes is "not in physical time" (see p. 506 above). But this claim can be denied only by denying the efficacy of the final causes within the occasion which Chappell would have Whitehead retain. The reason the act of becoming is not in physical time is because "physical time makes its appearance in the 'coordinate' analysis of the 'satisfaction'" (PR 434). There is no physical time until the act of becoming has terminated in its "satisfaction" and the entity is complete. Physical time does not extend to the process of becoming itself because "it is only the physical pole of the actual entity which is thus divisible. The mental pole is incurably one" (PR 436).

The *res vera,* in its character of concrete satisfaction, is divisible
into prehensions which concern its first temporal half and into
prehensions which concern its second temporal half. This divisibility is what constitutes its extensiveness. But this concern with a
temporal and spatial sub-region means that the datum of the
prehension in question is the actual world, objectified with the
perspective due to that sub-region. A prehension, however, ac-
quires subjective form, and this subjective form is only rendered
fully determinate by integration with conceptual prehensions
belonging to the mental pole of the *res vera.* The concrescence
is dominated by a subjective aim which essentially concerns the
creature as a final superject. This subjective aim is this subject
itself determining its own self-creation as one creature. Thus
the subjective aim does not share in this divisibility. If we confine
attention to prehensions concerned with the earlier half, their
subjective forms have arisen from nothing. For the subjective
aim which belongs to the whole is now excluded. Thus the evolution of subjective form could not be referred to any actuality.
The ontological principle has been violated. Something has
floated into the world from nowhere. (PR 108)

The process of concrescence is not divisible into parts with
temporal extension. If one abstracts the temporally divisible physical pole from the indivisible mental pole, that abstraction is
divisible into temporally extended parts; but "philosophy is the
critic of abstractions" (SMW 126).[4] The process by which an actual
occasion becomes is never merely physical. Since the mental pole is
indivisible and there is no actual occasion without a mental pole,
there is no occasion and, *a fortiori,* no physical pole to divide until
the process is complete. The process of concrescence can be understood only if the subjective aim is taken into account, and this
subjective aim can be understood only in terms of the theory of
final causation to which Chappell appeals.

Although, in his discussion of genetic division, Whitehead speaks
of earlier and later phases of the concrescence of an actual occasion
(e.g. PR 337) , these phases are not concrete entities. Their order is

---

4 This refers to the 1937 reprinting of the Macmillan edition (New York) of
Science and the Modern World, which corresponds to Chappell's citations, but not
to the 1925 edition. Sometime between 1925 and 1937 Macmillan changed the
pagination without indicating a new edition.

determined by consideration of efficient causal relations within the occasion in abstraction from the final causal influence of the subjective aim. In order to achieve temporal extension prior to the whole occasion, these phases would need to achieve actuality independently of this subjective aim. But to divorce a phase of the concrescence from its subjective aim is to eliminate its reason for being. It is not an appeal to Zenonian arguments which precludes temporal division here, but an appeal to the ontological principle, and although one might choose to deny the validity of the ontological principle, Chappell's attack does not take this form.

For Whitehead time and space are abstractions from concrete actual occasions. They have no reality apart from these occasions. The nature of time and space therefore depends upon the nature of actual occasions. Chappell's attempt to make temporal divisibility or indivisibility a problem distinct from the divisibility or indivisibility of concrete actual occasions denies this dependence. It suggests an independently existing spatio-temporal reality waiting to be filled by actual occasions and indifferent to the natural articulations of the concrete entities it will contain. Without such an independently existing extensive continuum, there is no basis for the distinction between temporal and nontemporal division. There is nothing to which the term "temporal" can refer except the actual occasion itself. Temporal division can only mean division of a process creative of time into processes creative of time, but, since nothing short of a whole actual occasion is creative of time, this would be the division of an actual entity "into processes that are themselves actual occasions" which Chappell allows Whitehead to retain.

While it is true that Whitehead speaks of an extensive continuum prior to its atomization by actual occasions as well as coordinate division of occasions which have become, this continuum "is not a fact prior to the world; it is the first determination of order—that is, of real potentiality—arising out of the general character of the world" (PR 103). The epochal character of time is no more threatened by the indefinite divisibility of a field of potentiality extending into the contemporary world and the future than it is compromised by the extensive character of the locus of the occasions which have come to be. The actual becoming of the occasion itself, which destroys the former continuity (PR 104) and creates the latter, is not, and cannot, be temporally divided. Having granted us the atomic character of Whitehead's actual entities, Chappell cannot deny us their temporal indivisibility—which is to say that he cannot deny us the epochal character of time.

III. Having demonstrated that Whitehead needs the epochal theory of time if he is to be allowed the atomic actual occasions Chappell is willing to grant him, we are now in a position to consider the arguments by which Chappell claims to render the epochal theory of time unintelligible. If we have indeed established Whitehead's need of the theory, these arguments do not concern "a minor appendage," as Chappell claims (525; 77), but strike "at the heart of Whitehead's philosophical position"; if Chappell's arguments are allowed to stand, and if our above discussion is sound, they render the whole Whiteheadian cosmology unintelligible. If the above discussion is sound, however, it also provides a basis for dismissing Chappell's arguments.

"The notion of a becoming that is not extensive is an odd notion at best," Chappell argues.

> Becoming is ordinarily thought of as a process that takes a certain time to occur and that extends, therefore, through a certain period of time. To say that a certain act or process of becoming is not extensive, however, would seem to imply that the becoming does not take any time to occur, that it is an act or process that extends through no time. The becoming may still occur at a time, i.e., at a certain instant, but only at that instant; hence it will be an instantaneous becoming. And it may be wondered how any act or process could be instantaneous, how it could (logically) take no time to occur. (521; 73-74)

It may indeed. But surely the question has been begged. Given our "ordinary" concept of becoming, Chappell argues, Whitehead's position would seem to imply that Whiteheadian becoming is instantaneous. He arrives at the conclusion by equating "not extensive" with "does not take any time to occur" and "does not take any time to occur" with "instantaneous." But it is precisely this kind of equation Whitehead is concerned to deny. Our ordinary understanding analyzes the process of becoming *ex post facto*, in terms of abstractions from what has—or more subtly, what might have—become. Within an abstract continuum we find little difficulty in constructing abstract instants. The act of becoming, however, is neither abstract nor instantaneous. Although it is not extensive, it "is not extensive, in the sense that it is divisible into earlier and later acts of becoming which correspond to the extensive divisibility of what has become" (PR 107). This has nothing to do with the instantaneous. The process does not take any time to occur only because time is created by its occurring. Before its occurring there is no time within which it can be instantaneous. There is no temporal extension—only the potentiality of temporal extension.[5] This potentiality is actualized by the occasion and the process or act of becoming by which it is actualized is an indivisible epoch (PR 103-104).

What has become is extensive; it has not taken time to become, but it has created time by becoming. The act of becoming fails to be extensive not because it does not endure but because its endurance is indivisible. Only if one presupposes that time is a mathematical continuum does "nonextensiveness" entail "instantaneousness." It is hardly surprising that the epochal theory of time is unintelligible if one presupposes that time is a mathematical continuum. Chappell has not rendered the epochal theory of time unintelligible; he has merely revealed its incompatibility with the ordinary language of temporal continuity.

"Even if this difficulty could be met and a sense found or provided for the notion of an instantaneous becoming or of an act or process that takes no time," however, Chappell assures us that "there is another difficulty to be faced."

> Concrescences, or acts of microscopic becoming, *take* no time in Whitehead's view; they could not do so and still be nonextensive. But they must occur *in* time; they must have dates even if they are without duration. But the events or occasions that are the products of such acts are also located in time. Indeed, they have both date and duration, according to Whitehead; the existence of each begins at one definite time, ceases at a later time, and extends through all the times between. But if both the act and the product of the act of becoming are in time, there must be some temporal relation between them. But what could this relation be?

---

[5] It is either this potentiality or a retrospective judgment which gives meaning to the word 'before' in the preceding sentence.

Certainly the act cannot be either altogether earlier or altogether later than the product; the act's date must coincide with one of the times through which the produced event exists. But we cannot say that the act is simultaneous with the event's time of cessation, for then we should have to grant that the event began before its act of becoming had occurred, which is absurd. Nor can we, for the same reason, say that the act occurs at any time later than that at which the event begins to be. Therefore we must conclude that an actual occasion's act of becoming coincides with the moment at which the occasion begins to exist. (521-522; 74)

Here the errors are compounded. The fundamental mistake is, again, the assumption that concrescences must occur in time. The act which is creative of time can hardly occur in time, nor is there any basis for claiming that the act's date must coincide with "one of the times," "times" clearly meaning "instants." To restrict dates to instants is to presuppose a mathematical continuum. Whitehead's occasions have no obligation to date themselves in terms of instants nor to fit themselves into a pre-existing continuum. There is no actual continuum pre-existing, nor any temporal elements, instantaneous or otherwise, until the occasion is complete. Chappell gives cogent reasons why the act cannot occur "at any time later than that at which the event begins to be." "Therefore," he assures us, "we must conclude that an actual occasion's act of becoming coincides with the moment at which the occasion begins to exist."

We must conclude nothing of the kind! We must conclude, rather, that the attempt to reduce an epoch to an instant fails. The act coincides with no moment at all, but with the entire temporal epoch which it creates by coming to be. Certainly the coincidence of "an actual occasion's act of becoming" with "the moment at which the occasion begins to exist . . . will not stand" (522; 74) ; but there is no reason whatsoever for Whitehead to suggest such a coincidence. Before he could even consider such a suggestion, he would have to restate it in the past tense, for this "moment" has meaning only after the occasion is complete. Then, having made Chappell's conclusion intelligible within the framework of his own position, he would agree that this coincidence is self-contradictory, and point out that it is also irrelevant. The act of becoming must be coincident with the epoch which it creates; there is no contradiction involved until one attempts to replace the epoch with an instant.

IV. Although we have referred specifically to Whitehead in considering Chappell's arguments, it is worth noting that an epochal theory of time need not be presented in Whiteheadian terms to meet these arguments. Any epochal theory of time escapes these arguments if it disallows instantaneous acts of becoming and defines the time during which the epoch is created as coincident with the duration of the epoch itself. As long as the theory bars the instantaneous from concrete existence, it has nothing to fear from the arguments presented so far, whether or not its temporal epochs exhibit all the complex properties of Whiteheadian occasions. Much of the remainder of Chappell's argument is vitiated by this question-begging assumption that an occasion must become at an instant. "An event or occasion is said to exist from time t_0 to time t_1," he argues, "because an act of becoming, the act whereby it comes into being, occurs at t_0" (522; 74) ; or, restated, "substituting 'happening' for 'becoming', . . . an actual occasion exists from t_0 to t_1 because it happens at t_0; the occasion, which is temporally extended, begins at the time of its happening, and its happening is itself instantaneous" (522; 75). We have already disposed of this argument, but in restating his conclusion in terms of "happening," Chappell gives the argument a twist which is of particular interest to us because it so clearly reveals the nature of the error involved in his analysis.

"What does it mean to say that an event or occasion comes into being?" Chappell asks. "*Things* come to be," he explains, referring us to J.J.C. Smart,[6] "but events *occur* or *happen*" (522; 74-75) . Turning to Smart, we read that

> Temporal facts are facts of before and after and of simultaneity. Now we may say, roughly, that it is events that are before and after one another or simultaneous with one another, and that events are happenings to things. Thus the traffic light changed from green to amber and then it changed from amber to red. Here are two happenings, and these happenings are changes of state of the traffic light. That is, *things* change, *events* happen. The traffic light changes, but the changing of the traffic light cannot

---

[6] "The River of Time," *Mind*, 58, No. 232 (October, 1949) , 483-94.

be said to change. To say that it does or does not change is to utter nonsense. Similarly, the traffic light neither does nor does not happen. We must also resist the temptation to misuse the word "become." The traffic light *was* green and *became* red, but the becoming red did not become. Events happen, things become, and things do not just become, they become something or other. "Become" is a transitive verb; if we start using it intransitively we can expect nothing but trouble. This is part of what is wrong with Whitehead's metaphysics. ("River of Time," 485-86)

Whether or not what Smart is analyzing here is ordinary language is quite irrelevant. What is relevant is that the language he is analyzing is clearly a language of enduring things with changing attributes. It is not surprising that Whitehead chooses to use language in a different way, for his metaphysics is not that which is implicit in the usage Smart is explicating. He explicitly rejects the metaphysics of enduring substances and changing attributes and has gone to great lengths to construct a vocabulary and usage adequate to the concrescence of actual occasions instead of the changing of things. For Whitehead the traffic light is a complex nexus of actual occasions and the particles of which it is composed are historic routes of occasions. If "this is part of what is wrong with Whitehead's metaphysics," it is not because Whitehead's usage or metaphysics is unintelligible, but because it differs from a usage and a metaphysics which Smart finds preferable. It is not within the scope of this paper to decide between a metaphysics of substance and a metaphysics of process; but it is important to make clear that what is in question here is not the proper use of words, nor even the intelligibility of a metaphysics of process properly stated. What is at issue is a choice between these two metaphysics.

"Broad (*Scientific Thought*, p. 68) agrees that events do not *change*," Smart notes, "but he says that they *become*, and by this he means that they *come into existence*."

Now this use of "become" is no more applicable to events than is the ordinary transitive use. Events do not come into existence; they occur or happen. "To happen" is not at all equivalent to "to come into existence" and we shall be led far astray if we use the two expressions as though they could be substituted for one another. We can say when the inauguration of a new republic occurred and we can say that the new republic came into existence then, but we cannot say that the inauguration came into existence. (486)

Whatever the status of Broad's events or Smart's, and whatever our proper usage when referring to the birth of nations, Whitehead's actual occasions do come into existence and Whitehead refers to this as their becoming. There is nothing unintelligible about such usage. If Chappell's project were to point out to us that Whitehead's actual occasions are not what Smart calls events, or that they are not what events are ordinarily taken to be, we would have no quarrel with him. But Chappell's claim is that Whitehead's position is unintelligible. Chappell demonstrates that if Smart's usage is substituted for Whitehead's, then Whitehead's position, stated in Smart's terms, is unintelligible. Of course it is; but this is hardly surprising, nor is it damaging to Whitehead.

"The products of acts of microscopic becoming *are* events or occasions according to Whitehead," Chappell asserts, "and are not things or objects. What Whitehead must mean, therefore, when he says that actual occasions become, is just that they happen; or at any rate he must allow that what 'becoming' signifies, when used of actual occasions, is no different from what we should ordinarily call 'happening' " (522; 75) . On the contrary, he must insist that the becoming of an actual occasion is quite different from what we should ordinarily call 'happening'. Actual occasions are not events if by 'event' one means what happens to things. Actual occasions certainly do not happen to things; there are no things in Whitehead's world until there are actual occasions. Things change; events occur; actual occasions come into being, and we shall refer to this as their becoming—whether or not it abuses grammar to do so.

Not only are actual occasions not events, as Chappell and Smart use the term; one must qualify the statement that they are not things. They are the closest things to substances we will find in Whitehead's philosophy, as he himself points out, and they have this character both as the process and as the product of their becoming. "The process of concrescence" satisfies "Spinoza's definition of substance, that it is *causa sui*" (PR 134-135) , and the "products of acts of microscopic becoming" (Chappell: 522; 75; quoted above) , being no longer processes, have a greater permanence and completeness than we attribute to everyday things.

The actual occasion must always be understood to be both process and product, both subject and superject (PR 43). Chappell allows Whitehead the distinction, but denies him its use. "Whitehead's doctrine is easily shown to be senseless," he assures us.

> For it requires us to distinguish the happening of an event or occasion from the event itself, or from the event's existence. But how does an event differ from its happening, or how does its existence do so? The answer is that it differs not at all. An event is not one thing and its happening another; happening adds nothing to an event, as being or existing adds nothing to a thing or object. An event, in short, *is* its happening; to be an event and to happen are one and the same. (522-523; 75)

Whitehead's doctrine is "easily shown to be senseless" only if one gives alien meanings to his terms—meanings which are no less alien if they should be proved to be more consistent with ordinary usage. Whitehead wishes to distinguish the concrete process of becoming from the product of this process, which continues to exist in an objective form even after the subjectivity of the process has terminated and the epoch it has created is no longer present, but is a part of the past of new occasions. There is a sense in which *the process* does not differ from its happening, though "happening" can't mean quite what it does for Chappell and Smart. Chappell does not identify events with processes here, however, but with "the products of acts of microscopic becoming" (p. 515 above). The product does not coincide with the happening in Chappell's sense or in ours. It has an objective immortality which the process does not share (PR 44). If the process is the happening and the product is the event, then "to be an event and to happen" are *not* "one and the same thing." To happen, *in these terms*, is to come into being; to be an event, *in these terms*, is to have come into being. These are the same thing only if "event" is given the meaning of process as well as of product, and this is precisely what Chappell has done.

Chappell tells us that

> Things (objects) come into being and then are; their becoming precedes their being; and although nothing can come into being without thereby being, it is at least conceivable that a thing should be and not have come to be. But if an event "becomes" by happening it also "is" by happening. Then if we choose, with Whitehead, to use "becoming" of events or actual occasions, we must ac-

knowledge, as Whitehead himself does not, that our meaning is no different from what it would have been had we used "being" instead. (523; 75)

If we paraphrase Chappell in terms appropriate to Whitehead rather than to Smart, we reach a different conclusion.

Actual occasions come into being and then, in a sense, continue to be, even after they no longer become; their becoming precedes their objective existence as products of this becoming. The product of becoming cannot come into being without thereby being; nor is it any more conceivable that it should be and not have come to be than that an event should be but not have happened. If we concern ourselves with the process of becoming, to say that this process "is" is to say that it "becomes"; but the "becoming" of an actual occasion must be carefully distinguished from the objective existence of the product of this becoming. We must acknowledge that Whitehead's meaning when he speaks of the becoming of an actual occasion is quite different from his meaning when he speaks of the objective existence of what has come into being. This distinction is not only legitimate, but necessary; these two meanings must not be confused.

The above passage from Chappell, for which we have provided this paraphrase, leads directly to the conclusion of his demonstration of the unintelligibility of the epochal theory of time.

Actual occasions, Whitehead says, are temporally extended, but their acts of becoming, which is to say their happenings, are not. But if an occasion is its happening, if it exists by happening, then Whitehead's doctrine ends in a contradiction: one self-same thing both is and is not extensive. And this is sufficient to show the unintelligibility of Whitehead's view. (523; 75)

On the contrary, there is no contradiction in asserting both that the products of acts of becoming are temporally extended and that the processes which produce these products are not. If one asserts that the process is its happening, he must also note that the product continues to exist as a datum for future processes after the process itself has already "happened" and no longer exists. There is no contradiction in saying that "one self-same thing both is and is not extensive" as long as one keeps these two distinct aspects of the thing carefully in mind. An actual occasion is extensive if one is referring to the product of concrescence; it is nonextensive if one is referring to the process itself. There is nothing unintelligible here.

V. If our arguments are sound, the epochal theory of time is both necessary, in the sense that it is entailed by what Chappell and I both accept as one of the "major and essential tenets in Whitehead's position," and tenable. It is necessary because the atomic nature of actual occasions entails the epochal theory of time. It is tenable because the contradictions Chappell puts before us disappear as soon as the atomic nature of actual occasions is taken seriously. Chappell has used a language unsuitable to Whitehead; he has presented Whitehead in terms which presuppose that Whitehead's own categories are inapplicable. Such a presentation necessarily begs the question. Once Chappell's arguments are restated in terms appropriate to the theory under consideration, the contradictions disappear. The epochal theory is quite intelligible; it is therefore worthy of serious consideration. Although it is probably clear to the reader that I prefer it to competing theories, why it is to be preferred is a subject for another paper—a paper which might turn to Zeno's paradoxes and to Chappell's discussions of them.[7]

DAVID A. SIPFLE

CARLETON COLLEGE

---

[7] Not only in "Whitehead's Theory of Becoming" (523-525; 75-77), but in "Time and Zeno's Arrow," *The Journal of Philosophy*, 59, No. 8 (April 12, 1962), 197-213.

THE MEANING OF TIME*

I. INTRODUCTION

Studies of time by scientists have often been concerned with the multifaceted problems of measuring time intervals in atomic, geophysical, biological, and astronomical contexts. It has been claimed that in addition to exhibiting measurable intervals, time is characterized by a *transiency* of the present, which has often been called 'flux' or 'passage'.

Indeed, it has been maintained that *'the passage of time . . . is the very essence of the concept'*.[1] I therefore wish to focus my concern with the meaning of time on the credentials which this transiency of the present can claim from the point of view of current physical theories.

In the common-sense view of the world, it is of the very essence of time that events occur now, or are past, or future. Furthermore, events are held to change with respect to belonging to the future or the present. Our commonplace use of tenses codifies our experience that any particular present is superseded by another whose event-content thereby 'comes into being'. It is this occurring *now* or coming into being of previously future events and their subsequent belonging to the past which is called 'becoming' or 'passage'. Thus, by involving reference to *present* occurrence, becoming involves more than mere occurrence at various serially ordered clock times. The past and the future can be characterized as respectively before and after the present. Hence I shall center my account of becoming on the status of the present or now as an attribute of events which is encountered in *perceptual* awareness.

---

*A Louis Clark Vanuxem Lecture, delivered at Princeton University on March 2, 1967. This paper includes a *revised* version of Chapter 1 of my *Modern Science and Zeno's Paradoxes* (Middletown: Wesleyan University Press, 1967); (2d ed.; London: Allen & Unwin, Ltd., 1968). An earlier revision appeared in *Essays in Honor of Carl G. Hempel* (Dordrecht, Holland: D. Reidel Publishing Co., 1969).

[1] G. J. Whitrow, *The Natural Philosophy of Time* (London: Thomas Nelson & Sons, Ltd., 1961), p. 88.

Granted that becoming is a prominent feature of our temporal awareness, I ask: *must* becoming therefore also be a feature of the order of physical events *independently* of our awareness of them, as the common-sense view supposes it to be? And if not, is there anything within physical theory *per se* to warrant this common-sense conclusion?

It is apparent that the becoming of physical events in our temporal awareness does not itself guarantee that becoming has a mind-independent physical status. Common-sense color attributes, for example, surely *appear* to be properties of physical objects independently of our awareness of them and are held to be such by common sense. And yet scientific theory tells us that they are mind-dependent qualities like sweet and sour are. Of course, if physical theory claims that, contrary to common sense, becoming is not a feature of the temporal order of physical events with respect to earlier and later, then a more comprehensive scientific and philosophical theory must take suitable cognizance of becoming as a conspicuous characteristic of our *temporal awareness* of both physical and mental events.

In this lecture, I aim to clarify the status of temporal becoming by dealing with each of the questions I posed. Clearly, an account of becoming which provides answers to these questions is *not* an analysis of what the common-sense man actually *means* when he says that a physical event belongs to the present, past, or future; instead, such an account sets forth how these ascriptions ought to be construed within the framework of a theory which would supplant the scientifically untutored view of common sense. That the common-sense view is indeed scientifically untutored is evident from the fact that *at a time t,* both of the following physical events qualify as occurring 'now' or 'belonging to the present' according to that view: (i) a stellar explosion that occurred several million years before time t but which is first seen on earth at time t, (ii) a lightning flash originating only a fraction of a second before t and observed at time t. If it be objected that present-day common-sense beliefs have *begun* to allow for the finitude of the speed of light, then I reply that they err at least to the extent of associating absolute simultaneity with the now.

The temporal relations of earlier (before) and later (after) can obtain between two physical events independently of the transient now and of any minds. On the other hand, the classification of events into past, present, and future, which is inherent to becoming, requires reference to the adverbial attribute now as well as to the relations of earlier and later. Hence the issue of the mind-dependence of becoming turns on the status of the adverbial attribute now. And to assert in this context that becoming is mind-dependent is *not* to assert that the obtaining of the relation of temporal precedence among physical events is mind-dependent. Nor is it to assert that the mere occurrence of events at various serially-ordered clock times is mind-dependent.

With these explicit understandings, I can state my thesis as follows: Becoming is mind-dependent because it is not an attribute of physical events per se but requires the occurrence of certain *conceptualized conscious experiences* of the occurrence of physical events. The doctrine that becoming is mind-dependent has been misnamed 'the theory of the block universe'. I shall therefore wish to dissociate the tenets of this doctrine both from serious misunderstandings by its critics and from the very misleading suggestions of the metaphors used by some of its exponents. Besides stating my positive reasons for asserting the mind-dependence of becoming, I shall defend this claim against the major objections which have been raised against it.

III.   THE DISTINCTION BETWEEN TEMPORAL BECOMING AND THE
ANISOTROPY OF TIME

In order to treat these various issues without risking serious confusions, we must sharply distinguish the following two questions: (i) do physical events *become* independently of any conceptualized awareness of their occurrence, and (ii) are there any kinds of physical or biological processes which are *irreversible* on the strength of the laws of nature and/or of *de facto* prevailing boundary conditions? I shall first state how these two questions have come to be identified and will then explain why it is indeed an error of consequence to identify them. The second of these questions, which pertains to irreversibility, is often formulated by asking whether the time of physics and biology has an 'arrow'. But this formulation of question (ii) can mislead by inviting misidentification of (ii) with (i). For the existence of an arrow is then misleadingly spoken of as constituting a 'one-way forward flow of time', but so also is becoming on the strength of being conceived as the forward 'movement' of the present. And this misidentification is then used to buttress the false belief that an affirmative answer to the question about irreversibility entails an affirmative answer to the question about becoming. To see why I claim that there is indeed a weighty misidentification here, let us first specify what is involved logically when we inquire into the existence of kinds of processes in nature which are irreversible.

If the system of world lines, each of which represents the career of a physical object, is to exhibit a one-dimensional temporal order, relations of simultaneity between spatially separated events are required to define world states. For our purposes, it will suffice to use the simultaneity criterion of some one local inertial frame of the special theory of relativity instead of resorting to the cosmic time of some cosmological model.

Assume now that the events belonging to *each* world-line are invariantly ordered with respect to all inertial systems by a *betweenness* relation having the following *formal* property of the spatial betweenness of the points on an Euclidean straight line: of any three elements, only one can be between the other two. This betweenness of the events is clearly temporal rather than spatial, since it *invariantly* relates the events belonging to each *individual world line* with respect to all inertial systems, while no such *spatial* betweenness obtains invariantly.[2] So long as the temporal betweenness of the world lines is formally Euclidean in the specified sense, any two events on one of them or any two world states can serve to define two time senses which are *ordinally* opposite to each other with respect to the assumed temporal betweenness relations.[3] And the members of the simultaneity-classes of events constituting one of

---

[2] For example, consider the events in the careers of human beings or of animals who *return* to a spatially fixed terrestrial habitat every so often. These events occur at space points on the earth which certainly do *not* exhibit the betweenness of the points on a Euclidean straight line.

[3] For details, cf. A. Grünbaum, "Space, Time and Falsifiability," Part I, Philosophy of Science, 37 (1970), 485–86 and 584–85. This treatment supersedes the earlier one in A. Grünbaum, *Philosophical Problems of Space and Time* (New York: Alfred A. Knopf, Inc., 1964), pp. 214–16, hereinafter cited as PPST.

these two opposite senses can then bear lower real number coordinates while those of the other sense can bear the higher coordinates.
It is immaterial at this stage which of the two opposite senses is
assigned the higher real numbers. All we require is that the real
number coordinatization reflect the temporal betweenness relations
among the events as follows: events which are temporally between
two given events E and E' must bear real number coordinates which
are numerically between the time coordinates of E and E'. Employing some one time coordinatization meeting this minimal requirement, we can use the locutions 'initial state', 'final state', 'before',
and 'after' on the basis of the magnitudes of the real number
coordinates, entirely without prejudice to whether there are irreversible kinds of processes.[4] By an 'irreversible process' (à la Planck)
we understand a process such that no counter-process is capable of
restoring the original *kind* of state of the system at another time.
Note that the temporal vocabulary used in this definition of what is
meant by an irreversible kind of process does *not* assume tacitly that
there *are* irreversible processes: as used here, the terms 'original
state', 'restore', and 'counter-process' presuppose only the coordinatization based on the assumed betweenness.

It has been charged that one is guilty of an illicit spatialization
of time if one speaks of temporal betweenness while still leaving it
open whether there are irreversible kinds of processes. But this
charge overlooks that the *formal* property of the betweenness on the
Euclidean line which I invoked is abstract and, as such, neither
spatial nor temporal. And the meaningful attribution of this formal
property to the betweenness relation among the events belonging to
each world line without any assumption of irreversibility is therefore *not* any kind of illicit spatialization of time. As well say that
since temporal betweenness does have this abstract property, the

---

[4] This noncommittal character of the term 'initial state' seems to have been
recognized by O. Costa de Beauregard in one part of his paper entitled 'Irreversibility Problems', *Logic, Methodology and Philosophy of Science*, Proceedings of
the 1964 International Congress. Y. Bar-Hillel, ed. (Amsterdam: North-Holland
Publishing Co., 1965) , p. 327. But when discussing my criticism of Hans Reichenbach's account of irreversibility (PPST, pp. 261–63) , Costa de Beauregard (*ibid.*, p.
331) overlooks that my criticism invokes initial states in only the noncommittal
sense set forth above.

ascription of the latter to the betweenness among the points on a line of space is a temporalization of space![5]

Thus the assumption that the events belonging to each world line are invariantly ordered by an abstractly Euclidean relation of temporal betweenness does not entail the existence of irreversible kinds of processes, but allows every kind of process to be reversible.[6] If there are irreversible processes, then the two ordinally opposite time senses are indeed *further* distinguished structurally as follows: there are certain kinds of sequences of states of systems specified in the order of increasing time coordinates such that these same kinds of sequences do *not* likewise exist in the order of decreasing time coordinates. Or, equivalently, the existence of irreversible processes *structurally* distinguishes the two opposite time senses as follows: there are certain kinds of sequences of states of systems specified in the order of *decreasing* time coordinates such that these same kinds of sequences do *not* likewise obtain in the order of increasing time coordinates. Accordingly, if there are irreversible kinds of processes,

---

[5] Thus, it is erroneous to maintain, as Milic Capek does, that the distinction between temporal betweenness and irreversibility is 'fallacious' in virtue of being 'based on the superficial and deceptive analogy of "the course of time" with a geometrical line', *The Philosophical Impact of Contemporary Physics* (Princeton: D. Van Nostrand Co., Inc., 1961), p. 349; see also pp. 347 and 355. If Capek's condemnation of this distinction were correct, the following fundamental question of theoretical physics could not even be intelligibly and legitimately asked: Are the *prima facie* irreversible processes known to us indeed irreversible, and, if so, on the strength of what laws and/or boundary conditions are they so? For this question is predicated on the very distinction which Capek rejects as 'fallacious'. By the same token, Capek errs (*ibid.*, p. 355) in saying that when Reichenbach characterizes entropically counterdirected epochs as 'succeeding each other', then irreversibility 'creeps in' along with the asymmetrical relations of before and after. For all that he needs to assume here to speak of 'before' and 'after' is a time coordinatization which reflects the assumed kind of betweenness and simultaneity.

[6] On the basis of a highly equivocal use of the term 'irreversible', M. Capek, *ibid.*, pp. 166—67 and 344—45 has claimed incorrectly that the account of the space-time properties of world lines given by the special theory of relativity entails the irreversibility of physical processes represented by world lines. He writes: 'The world lines, which by definition are constituted by a succession of isotopic events, are *irreversible* in all systems of reference' (*ibid.*, p. 167) and 'the relativistic universe is dynamically constituted by the network of causal lines *each of which is irreversible*; . . . this irreversibility is a topological invariant' (*ibid.*, pp. 344–345). But Capek fails to distinguish between (1) the *non-inversion or invariance of time-order as between different Galilean frames* which the Lorentz-transformation

then time is *anisotropic*.[7] When physicists say with Eddington that time has an 'arrow', it is this anisotropy to which they are referring metaphorically. Specifically, the spatial opposition between the head and the tail of the arrow represents the structural anisotropy of time.

Note that we were able to characterize a process as irreversible and time as anisotropic without any explicit or tacit reliance on the transient now or on tenses of past, present, and future.[8] By the same token, we are able to assert metaphorically that time has an 'arrow' without any covert or outright reference to events as occurring *now*, happening at present, or coming into being. Nonetheless, the anisotropy of time symbolized by the arrow has been falsely equated in the literature with the transiency of the now or becoming of events via the following steps of reasoning: (1) the becoming

---

equations assert in the case of causally connectible events, and (2) the irreversibility of processes represented by world lines in the standard sense of the *non-restorability* of the same kind of state in any frame. Having applied the term 'irreversibility' to (1) no less than to (2) after failing to distinguish them, Capek feels entitled to infer that the Lorentz transformations attribute irreversibility within any one frame to processes depicted by world lines, just because these transformations assert the invariance of time order on the world lines as between different frames. That the Lorentz equations do not disallow the reversibility of physical processes becomes clear upon making each of the *two* replacements $t \to -t$ and $t' \to -t'$ in them: these replacements issue in the same set of equations except for the sign of the velocity term in each of the numerators, i.e., they merely reverse the direction of the motion. Therefore, these two replacements do *not* involve any violation of the theory's time-order invariance as between different frames S and S'. By contrast, different equations exhibiting a violation of time-order invariance on the world lines would be obtained by replacing *only* one of the two variables t and t' by its negative counterpart in the Lorentz equations.

Furthermore, the transformation $t \leftrightarrow -t$ is a topological one, but it clearly does not preserve the time sequence relations on a time-like world line. Hence it is unsound for Capek to characterize the time-order invariance on time-like world lines as "topological."

[7] For a discussion of the various kinds of irreversible processes which make for the anisotropy of time and furnish specified criteria for the relations of temporal precedence and succession, see Costa de Beauregard, *op. cit.*, p. 327; and A. Grünbaum, PPST, Ch. 8, and 'The Anisotropy of Time', in *The Nature of Time* (ed. by D. L. Schumacher and T. Gold) (Ithaca: Cornell University Press, 1967), pp. 149–86.

[8] Some have questioned the possibility of stating what specific physical events do occur in point of fact at particular clock times without covert appeal to the transient now. Cf. Hermann Weyl, *Philosophy of Mathematics and Natural*

of events is described by the kinematic metaphor 'the flow of time' and is conceived as a *shifting* of the now which *singles out the future direction of time* as the sense of its 'advance', and (2) although the physicist's arrow does not involve the transient now, his assertion that there is an arrow of time is taken to be equivalent to the claim that there is a *flow* of time in the direction of the future; this is done by attending to the head of the arrow *to the neglect of its tail* and identifying the former with the direction of 'advance' of the now. The physicist's assertion that time has an 'arrow' discerningly codifies the empirical fact that the two ordinally opposite time senses are *structurally different* in specified respects. But in thus codifying this empirical fact, the physicist does *not* invoke the transient now to single out one of the two time senses as preferred over the other. By contrast, the claim that the present or now shifts in the direction of the future does invoke the transient now to single out one of the two time senses and—as we are about to see—is a mere truism like 'All bachelors are males'. Specifically, the terms 'shift' or 'flow' are used in their literal kinematic senses in such a way that the *spatial* direction of a shift or flow is specified by where the shifting object is at *later* times. Hence when we speak metaphorically of the now as 'shifting' temporally in a particular *temporal* direction, it is then simply a matter of

---

Science (Princeton: Princeton University Press, 1949), p. 75. In their view, any physical description will employ a time coordinatization, and any such coordinatization must ostensively invoke the now to designate at least one state as, say, the origin of the time coordinates. But I do not see a genuine difficulty here for three reasons. Firstly, it is not clear that the designation of the birth of Jesus, for example, as the origin of time coordinates tacitly makes logically indispensable use of the now or of tenses in virtue of making use of a proper name. Secondly, in some cosmological models of the universe, an origin of time coordinates can clearly be designated non-ostensively: in the 'big bang' model, the big bang itself can be designated uniquely and *non*-ostensively as the one state having no temporal predecessor. And thirdly, any two descriptions of the world which differ only in the choice of the origin of time coordinates while employing the same time metric and time topology are equivalent with respect to their factual *physical* content. Thus such descriptions differ only in regard to the way in which they numerically name or label particular simultaneity classes of events. Hence, let us grant for argument's sake that tacit use of the now or of tenses is logically indispensable to designating the origin of any one particular time coordinatization. Even if this is granted, it does *not* follow that past, present, and future have a mind-independent status in the temporal structure of the physical world.

definition that the now shifts or advances in the direction of the future. For this declaration tells us no more than that the nows corresponding to later times are later than those corresponding to earlier ones, which is just as uninformative as the truism that the earlier nows precede the later ones.[9]

It is now apparent that to assert the existence of irreversible processes in the sense of physical theory by means of the metaphor of the arrow does not entail at all that there is a mind-independent becoming of physical events as such. Hence those wishing to assert that becoming is independent of mind cannot rest this claim on the anisotropy of physical time.

Being only a tautology, the kinematic metaphor of time flowing in the direction of the future does not itself render any empirical fact about the time of our experience. But the role played by the present in becoming is a feature of the experienced world codified by common-sense time in the following informative sense: to each of a great diversity of events which are ordered with respect to earlier and later by physical clocks, there corresponds one or more particular experiences of the event as occurring *now*. Hence we shall say that our experience exhibits a *diversity of 'now-contents'* of awareness which are temporally ordered with respect to each of the relations earlier and later. Thus, it is a significant feature of the experienced world codified by common-sense time that there is a sheer diversity of nows, and in that sheer diversity the role of the future is no greater than that of the past. In this *directionally-*

---

[9] The claim that the now advances in the direction of the future is a truism as regards both the correspondence between nows and *physically* later clock times and their correspondence with psychologically (introspectively) later contents of awareness. What is *not* a truism, however, is that the *introspectively* later nows are *temporally correlated* with states of our physical environment that are later as per criteria furnished by irreversible physical processes. This latter correlation depends for its obtaining on the laws governing the physical and neural processes necessary for the *mental* accumulation of memories and for the registry of information *in awareness*. (For an account of some of the relevant laws, see A. Grünbaum, PPST, Ch. 9, Secs. A and B.) Having exhibited the aforementioned truisms as such and having noted the role of the empirical laws just mentioned, I believe to have answered Costa de Beauregard's complaint (in 'Irreversibility Problems', *op. cit.*, p. 337) that 'stressing that the arrows of entropy and information increase are parallel to each other is *not* proving that the flow of subjectivistic time has to follow the arrows!'

neutral sense, therefore, it is informative to say that there is a *transiency* of the now or a coming-into-being of different events. And, of course, in the context of the respective relations of earlier and later, this flux of the present makes for events being past and future.

In order to deal with the issue of the mind-dependence of becoming, I wish to forestall misunderstandings that can arise from uses of the terms 'become', and 'come into being' in senses which are *tenseless*. These senses do not involve belonging to the present or occurring now as understood in tensed discourse, and I must emphasize strongly that my thesis of the mind-dependence of becoming pertains only to the *tensed* variety of becoming. Examples of tenseless uses of the terms 'come into being', 'become', and 'now' are the following: (1) A child *comes into being* as a legal entity the moment it is conceived biologically. What is meant by this possibly false assertion is that for legal purposes, the career of a child *begins* (tenselessly) at the moment at which the ovum is (tenselessly) fertilized. (2) If gunpowder is suitably ignited at any particular time t, an explosion *comes into being* at that time t. The species of coming into being meant here involves a common-sense event which is here asserted to *occur tenselessly at time t*. (3). When heated to a suitable temperature, a piece of iron *becomes* red. Clearly, this sentence asserts that after a piece of iron is or has been (tenselessly) suitably heated, it is (tenselessly) red for an unspecified time interval. (4) In Minkowski's two-dimensional spatial representation of the space-time of special relativity, the event shown by the origin-point is called the 'Here-NOW', and correlatively certain event classes in the diagram are respectively called 'Absolute PAST' and 'Absolute FUTURE'; but Minkowski's 'Here-NOW' denotes an arbitrarily chosen event of reference which can be chosen *once and for all* and continues to qualify as 'now' at various times independently of when the diagram is used. Hence there is no transiency of the now in the relativistic scheme depicted by Minkowski, and his absolute past and absolute future are simply absolutely earlier and absolutely later than the arbitrarily chosen fixed reference event called 'Here-NOW'.[10] Accordingly, we must be

---

[10] A very illuminating account of the logical relations of Minkowski's language to tensed discourse is given by Wilfrid Sellars in 'Time and the World Order',

mindful that there are tenseless senses of the words 'becoming' and 'now'.

But conversely, we must realize that some important *seemingly* tenseless uses of the terms 'to exist', 'to occur', 'to be actual', and 'to have being or reality' are in fact laden with the present tense. Specifically, all of these terms are often used in the sense of to occur NOW. And by tacitly making the *nowness* of an event a necessary condition for its occurrence, existence, or reality, philosophers have argued fallaciously as follows. They first assert that the universe can be held to exist only to the extent that there are present events. Note that this either asserts that only present events exist now (which is trivial) or it is false. They then invoke the correct premiss that the existence of the physical universe is not mind-dependent and conclude (from the first assertion) that being present, occurring now, or becoming is *independent* of mind or awareness. Thus, Thomas Hobbes wrote: 'The present only has a being in nature; things past have a being in the memory only, but things to come have no being at all, the future being but a fiction of the mind....'[11] When declaring here that only present events or present memories of past events 'have being', Hobbes *appears* to be appealing to a sense of 'to have being' or of 'to exist' which is *logically independent* of the concept of existing-NOW. But his claim depends for its plausibility on the tacit invocation of *present* occurrence as a logically necessary condition for having being or existing. Once this fact is recognized, his claim that 'the present only has a being in nature' is seen to be the mere tautology that 'only what exists now does indeed exist now'. And by his covert appeal to the irresistible conviction carried by this triviality, he makes plausible the utterly unfounded conclusion that nature can be held to exist only to the extent that there are *present* events and *present* memories of past events. Clearly the fact that an event does not occur now does not justify the conclusion that it does not occur at some time or other.

IV.   THE MIND-DEPENDENCE OF BECOMING

Being cogizant of these logical pitfalls, we can turn to the following

---

Minnesota Studies in the Philosophy of Science, Vol. III (ed. by H. Feigl and G. Maxwell) (Minneapolis: University of Minnesota Press, 1962), p. 571.

11 Quoted from G. J. Whitrow, *op. cit.,* pp. 129–30.

important question: if a physical event occurs *now* (at present, in the present), what attribute or relation of its occurrence can warrantedly be held to qualify it as such?

In asking this question, I am being mindful of the following fact: if at a given clock time t_0 it is true to say of a particular event E that it is occurring now or happening at present, then this claim could not also be truly made at all other clock times $t \neq t_0$. And hence we must distinguish the tensed assertion of *present* occurrence from the tenseless assertion that the event E occurs at the time t_0: namely, the latter tenseless assertion, if true at all, can truly be made at all times t other than t_0 no less than at the time t_0. By the same token we must guard against identifying the tensed assertion, made at some particular time t_0, that the event E happens *at present* with the tenseless assertion made at *any* time t, that the event E occurs or 'is present' at time t_0. And similarly for the distinction between the tensed senses of being past or being future, on the one hand, and the tenseless senses of being *past at time* t_0 or being future at time t_0, on the other. To be future at time t_0 just means to be later than t_0, which is a tenseless relation. Thus our question is: what *over and above its otherwise tenseless occurrence at a certain clock time t*, in fact at a time t characterizes a physical event as *now* or as belonging to the present? It will be well remembered from the *Introduction* why my construal of this question does *not* call for an analysis of the common-sense meaning of 'now' or of 'belonging to the present' but for a critical assessment of the status which common sense attributes to the present.[12] Given this construal of the question, my reply to it is: what qualifies a physical event at a time t as belonging to the present or as now is *not* some physical attribute of the event or some relation it sustains to other *purely physical* events. Instead what is *necessary* so to qualify the event is that at the time t at least one human or other *mind-possessing* organism M is conceptually aware of experiencing at that time either the event itself or another event simultaneous with it in M's reference

---

[12] For a searching treatment of the ramifications of the contrast pertinent here, see Wilfrid Sellars, 'Philosophy and the Scientific Image of Man', *Frontiers of Science and Philosophy* (ed. by Robert G. Colodny) (Pittsburgh: University of Pittsburgh Press, 1962), pp. 35–78.

frame.[13] And that awareness does not, in general, comprise information concerning the date and numerical clock time of the occurrence of the event. What then is the content of M's conceptual awareness at time t that he *is experiencing* a certain event *at that time*? M's experience of the event at time t is coupled with an awareness of the temporal coincidence of his experience of the event with a state of *knowing* that he has that experience at all. In other words, M experiences the event at t *and* knows that he is experiencing it. Thus, presentness or nowness of an event requires conceptual awareness of the presentational immediacy of either the experience of the event or, if the event is itself *unperceived*, of the *experience* of another event simultaneous with it. For example, if I just hear a noise at a time t, then the noise does not qualify at t as *now* unless at t I am judgmentally aware of the fact of my hearing it at all and of the temporal coincidence of the hearing with that awareness.[14] If

---

[13] It will be noted that I speak here of the dependence of nowness on an organism M which is mind-possessing in the sense of having conceptualized or judgmental awareness, as contrasted with mere sentiency. Since biological organisms other than man (e.g., extra-terrestrial ones) may be mind-possessing in this sense, it would be unwarrantedly restrictive to speak of the mind-dependence of nowness as its 'anthropocentricity'. Indeed, it might be that conceptualized awareness turns out not to require a *biochemical* substratum but can also inhere in a suitably complex 'hardware' computer. That a physical substratum of some kind is required would seem to be abundantly supported by the known dependence of the content and very existence of consciousness in man on the adequate functioning of the human body.

[14] The distinction pertinent here between the *mere* hearing of something and judgmental awareness that it is being heard is well stated by Roderick Chisholm as follows: 'We may say of a man simply that he observes a cat on the roof. Or we may say of him that he observes *that* a cat is on the roof. In the second case, the verb "observe" takes a "that"-clause, a propositional clause as its grammatical object. We may distinguish, therefore, between a "propositional" and a "nonpropositional" use of the term "observe", and we may make an analogous distinction for "perceive", "see", "hear", and "feel".

'If we take the verb "observe" propositionally, saying of the man that he observes that a cat is on the roof, or that he observes a cat to be on the roof, then we may also say of him that he *knows* that a cat is on the roof; for in the propositional sense of "observe", observation may be said to imply knowledge. But if we take the verb non-propositionally, saying of the man only that he observes a cat which is on the roof, then what we say will not imply that he knows that there is a cat on the roof. For a man may be said to observe a cat, to see a cat, or hear a cat, in the nonpropositional sense of these terms, without his knowing that a cat is

the event at the time t is itself a mental event (e.g., a pain), then there is no distinction between the event and our experience of it. With this understanding, I claim that the nowness at a time t of either a physical or a mental event requires that there be an *experience* of the event or of another event simultaneous with it which satisfies the specified requirements. And by satisfying these requirements, the *experience* of a physical event qualifies at the time t as occurring *now*. Thus, the fulfillment of the stated requirements by the *experience* of an event at time t is also *sufficient* for the nowness of that *experience* at the time t. But the mere fact that the experience of a physical event qualifies as now at a clock time t allows that in point of physical fact the physical event itself occurred millions of years before t, as in the case of now seeing an explosion of a star millions of light years away. Hence, the mere presentness of the experience of a physical event at a time t does *not* warrant the conclusion that the clock time of the event is t or some *particular* time before t. Indeed, the occurrence of an external physical event E can never be simultaneous in any inertial system with the direct perceptual registration of E by a conceptualizing organism. Hence if E is presently experienced as happening at some particular clock time t, then there is no inertial system in which E occurs at that *same* clock time t. Of course, for *some* practical purposes of daily life, a nearby terrestrial flash in the sky can be held to be simultaneous with someone's experience of it with impunity, whereas the remote stellar explosion of a supernova or an eclipse of the sun, for example, may not. But this kind of practical impunity of common-sense perceptual judgments of the presentness of physical events cannot detract from their scientific falsity. And hence I do not regard it as incumbent upon myself to furnish a philosophical account of the status of nowness which is compatible with the now-verdicts of common sense. In particular, I would scarcely countenance making the nowness of the *experience* of a physical event *sufficient* for the nowness of the event, and even informed common sense might balk at this in cases such as a stellar explosion. But all that is essential to my thesis of mind-dependence

---

what he is observing, or seeing, or hearing. "It was not until the following day that I found out that what I saw was only a cat"'. *Theory of Knowledge* (Englewood Cliffs, New Jersey: Prentice-Hall, 1966), p. 10. I am indebted to Richard Gale for this reference.

is that the nowness of the *experience* of at least one member of the simultaneity class to which an event *E* belongs is *necessary* for the nowness of the event *E* itself. And hence my thesis would allow a compromise with common sense to the following extent: allowing ascriptions of nowness to those physical events which have the very vague relational property of occurring only 'slightly earlier' than someone's appropriate experience of them.

Note several crucial commentaries on my characterization of the now:

(1) My characterization of *present* happening or occurring *now* is intended to *deny* that belonging to the present is a physical attribute of a physical event *E* which is *independent* of any *judgmental awareness* of the occurrence of either *E* itself or of another event simultaneous with it. But I am *not* offering any kind of *definition* of the adverbial attribute now, which belongs to the conceptual framework of tensed discourse, solely in terms of attributes and relations drawn from the tenseless (Minkowskian) framework of temporal discourse familiar from physics. In particular, I avowedly invoked the present tense when I made the nowness of an event *E* at time *t* dependent on someone's knowing at *t* that he *is experiencing E.* And this is tantamount to someone's judging at *t*: I am experiencing *E now.* But this formulation is *non*viciously circular. For it serves to articulate the mind-dependence of nowness, *not* to claim erroneously that nowness has been eliminated by explicit definition in favor of tenseless temporal attributes or relations. In fact, I am very much less concerned with the adequacy of the specifics of my characterization than with its thesis of mind-dependence.

(2) It makes the nowness of an event at time *t* depend on the existence of conceptualized awareness that an experience of the event or of an event simultaneous with it is being had at *t*, and points out the insufficiency of the mere having of the experience. Suppose that at time *t* I express such conceptualized awareness in a linguistic utterance, the utterance being quasi-simultaneous with the experience of the event. Then the utterance satisfies the condition necessary for the *present* occurrence of the experienced event.[15]

---

15 The judgmental awareness which I claim to be essential to an event's qualifying as now may, of course, be expressed by a linguistic utterance, but it

(3) *In the first instance,* it is only an experience (i.e., a mental event) which can ever qualify as occurring now, and moreover a mental event (e.g., a pain) must meet the specified awareness requirements in order to qualify. A *physical* event like an explosion can qualify as now at some time *t* only *derivatively* in one of the following two ways: (a) it is necessary that someone's *experience* of the physical event does so qualify, or (b) if unperceived, the physical event must be simultaneous with another physical event that does so qualify in the derivative sense indicated under (a). For the sake of brevity, I shall refer to this complex state of affairs by saying that physical events belonging to regions of space-time wholly devoid of conceptualizing percipients at no time qualify as occurring now and hence as such do not become.

(4) My characterization of the now is narrow enough to exclude past and future events: It is to be understood here that the *reliving* or anticipation of an event, however vivid it may be, is *not* to be misleadingly called 'having an experience' of the event when my characterization of the now is applied to an experience.

My claim that nowness is mind-dependent does not assert at all that the nowness of an event is arbitrary. On the contrary, it follows from my account that it is not at all arbitrary what event or events qualify as being *now* at any given time *t*: to this extent, my account accords with common sense. But I repudiate much of what common sense conceives to be the status of the now. Thus, when I wonder in thought (which I *may* convey by means of an interrogative verbal utterance) whether it is 3 P.M. Eastern Standard Time now, I am asking myself the following: Is the particular percept of which I am now aware when asking this question a member of the simultaneity class of events which qualify as occurring at 3 P.M., E.S.T. on this

---

clearly need not be so expressed. I therefore consider an account of nowness which is *confined* to utterances as inadequate. Such an overly restrictive account is given in J. J. C. Smart's otherwise illuminating defense of the anthropocentricity of tense, *Philosophy and Scientific Realism* (London: Routledge & Kegan Paul, 1963), Chapter VII. But this undue restrictiveness is quite inessential to his thesis of the anthropocentricity of nowness. And the non-restrictive treatment which I am advocating in its stead would obviate his having to rest his case on (1) denying that 'this utterance' can be analyzed as 'the utterance which is *now*', and (2) insisting that 'now' must be elucidated in terms of 'this utterance' (*ibid.*, pp. 139-40).

particular day? And when I wonder in thought about what is happening now, I am asking the question: What events of which I am not aware are simultaneous with the particular now-percept of which I *am* aware upon asking this question?

That the nowness attribute of an occurrence, when ascribed non-arbitrarily to an event, is inherently mind-dependent seems to me to emerge from a consideration of the kind of information which the judgment 'It is 3 P.M., E.S.T. now' can be warrantedly held to convey. Clearly such a judgment is informative, unlike the judgment 'All bachelors are males'. But if the word 'now' in the informative temporal judgment does not involve reference to a particular content of conceptualized awareness or to the linguistic utterance which renders it at the time, then there would seem to be nothing left for it to designate other than either the time of the events already identified as occurring at 3 P.M., E.S.T. or the time of those identified as occurring at some other time. In the former case, the initially informative temporal judgment 'It is 3 P.M., E.S.T. now' turns into the utter triviality that the events of 3 P.M., E.S.T. occur at 3 P.M., E.S.T.! And in the latter case, the initially informative judgment, if false in point of fact, becomes self-contradictory like 'No bachelors are males'.

What of the retort to this objection that independently of being perceived physical events themselves possess an unanalyzable property of nowness (i.e., presentness) at their respective times of occurrence over and above merely occurring at these clock times? I find this retort wholly unavailing for several reasons as follows: (1) It must construe the assertion 'It is 3 P.M., E.S.T. now' as claiming *non-trivially* that when the clock strikes 3 P.M. on the day in question, this clock event and all of the events simultaneous with it intrinsically have the unanalyzable property of nowness or presentness. But I am totally at a loss to see that anything non-trivial can possibly be asserted by the claim that at 3 P.M. nowness (presentness) inheres in the events of 3 P.M. For all I am able to discern here is that the events of 3 P.M. are indeed those of 3 P.M. on the day in question! (2) It seems to me of decisive significance that nowness, in the sense associated with becoming, plays no role as a property of physical events themselves in any of the extant theories of physics. There have been allegations in the literature (most recently in H. A. C. Dobbs, 'The "Present" in Physics', *British*

Journal for the Philosophy of Science **19** (1968–1969) , 317–24) that such branches of statistical physics as meteorology and indeterministic quantum mechanics implicitly assert the existence of a physical counterpart to the human sense of the present. But both below (§V) and elsewhere (in my Reply to Dobbs in the *British Journal for the Philosophy of Science,* **20** (1969) , 145–53) , I argue that these allegations are mistaken. Hence I maintain that if nowness were a mind-*ind*ependent property of physical events themselves, it would be very strange indeed that it could be omitted *as such* from all extant physical theories *without detriment to their explanatory success.* And I hold with Reichenbach[16] that 'if there is Becoming [independently of awareness] the physicist must know it'. (3) As we shall have occasion to note near the end of Section V, the thesis that nowness is *not* mind-dependent poses a serious perplexity pointed out by J. J. C. Smart, and the defenders of the thesis have not even been able to hint how they might resolve that perplexity without utterly trivializing their thesis.

The claim that an event can be now (present) only upon either being experienced or being simultaneous with a suitably experienced event accords fully, of course, with the common-sense view that there is no more than one time at which a particular event is present and that this time cannot be chosen arbitrarily. But if an event is ever experienced at all such that there is simultaneous awareness of the fact of that experience, then there exists a time at which the event does qualify as being now provided that the event occurs only 'slightly earlier' than the experience of it.

The relation of the conception of becoming espoused here to that of common sense may be likened to the relation of relativity physics to Newtonian physics. My account of nowness as mind-dependent disavows rather than vindicates the common-sense view of its status. Similarly, relativity physics entails the falsity of the results of its predecessor. Though Newtonian physics thus cannot be reduced to relativity physics (in the technical sense of reducing one theory to another) , the latter enables us to see why the former works as well as it does in the domain of low velocities: relativity theory shows (via a comparison of the Lorentz and Galilean trans-

---

16 Hans Reichenbach, *The Direction of Time* (Berkeley: University of California Press, 1956) , p. 16.

formations) that the observational results of the Newtonian theory in that domain are sufficiently correct numerically for some practical purposes. In an analogous manner, my account of nowness enables us to see why the common-sense concept of becoming can function as it does in serving the pragmatic needs of daily life.

A *now-content* of awareness can comprise awareness that one event is later than or succeeds another, as in the following examples: (1) When I perceive the 'tick-tock' of a clock, the 'tick' is not yet part of my past when I hear the 'tock'.[17] As William James and Hans Driesch have noted, melody awareness is another such case of quasi-instantaneous awareness of succession.[18] (2) Memory states are contained in now-contents when we have awareness of other events as being earlier than the event of our awareness of them. (3) A now-content can comprise an envisionment of an event as being later than its ideational anticipation.

V. CRITIQUE OF OBJECTIONS TO
THE MIND-DEPENDENCE OF BECOMING

Before dealing with some interesting objections to the thesis of the mind-dependence of becoming, I wish to dispose of some of the caricatures of that thesis with which the literature has been rife under the misnomer of 'the theory of the block universe'. The worst of these is the allegation that the thesis asserts the timelessness of the universe and espouses, in M. Capek's words, the 'preposterous view . . . that . . . time is merely a huge and chronic [sic!] hallucination of the human mind'.[19] But even the most misleading of the spatial metaphors that have been used by the defenders of the mind-dependence thesis do not warrant the inference that the thesis denies the objectivity of the so-called 'time-like separation' of events known from the theory of relativity. To assert that nowness, and thereby, pastness and futurity are mind-dependent is surely *not* to assert that the earlier-later relations between the events of a world line are mind-dependent, let alone hallucinatory.

The mind-dependence thesis does deny that physical events

---

[17] Paul Fraisse, *The Psychology of Time* (London: Eyre & Spottiswoode, 1964), p. 73.

[18] A. Grünbaum, PPST, p. 325.

[19] M. Capek, *op. cit.*, p. 337.

themselves happen in the tensed sense of coming into being apart from anyone's awareness of them. But this thesis clearly avows that physical events do happen independently of any mind in the tenseless sense of merely occurring at certain clock times in the context of objective relations of earlier and later. Thus it is a travesty to equate the objective *becominglessness* of physical events asserted by the thesis with a claim of *timelessness*. In this way the thesis of mind-dependence is misrepresented as entailing that all events happen simultaneously or form a '*totum simul*'.[20] But it is an egregious blunder to think that if the time of physics lacks *passage* in the sense of there not being a transient now, then physical events cannot be temporally separated but must all be simultaneous.

A typical example of such a misconstrual of Weyl's and Einstein's denial of physical passage is given by supposing them to have claimed 'that the world is like a film strip: the photographs *are already there* and are merely being exhibited to us'.[21] But when photographs of a film strip 'are already there', they all exist now and hence *simultaneously*. Therefore it is wrong to identify Weyl's denial of physical becoming with the pseudo-image of the 'block universe' and then to charge his denial with entailing the absurdity that all events are simultaneous. Thus Whitrow says erroneously: 'the theory of "the block universe" . . . implies that past (and future) events co-exist with those that are present'.[22] We shall see in Section VI that a corresponding error vitiates the allegation that determinism entails the absurd contemporaneity of all events. And it simply begs the question to declare in this context that 'the *passage of time* . . . is the very essence of the concept'.[23] For the undeniable fact that passage in the sense of transiency of the now is integral to the common-sense concept of time may show only that, in this respect, this concept is anthropocentric.

The becomingless physical world of the Minkowski representa-

---

[20] On the basis of such a misunderstanding, M. Capek incorrectly charges the thesis with a 'spatialization of time' in which 'successive moments already *coexist*' (*ibid.*, pp. 160–63) and in which 'the universe with its whole history is conceived as a single huge and timeless bloc, given at once' (*ibid.*, p. 163) . See also p. 355.

[21] G. J. Whitrow, *op. cit.*, p. 228 (my italics) . For a criticism of another such misconstrual, see A. Grünbaum, PPST, pp. 327–28.

[22] G. J. Whitrow, *ibid.*, p. 88.

[23] *Ibid.*, pp. 227–28.

tion is viewed *sub specie aeternitatis* in that representation in the sense that the relativistic account of time represented by it makes no reference to the particular times of anyone's *now*-perspectives. And, as J. J. C. Smart observed, 'the tenseless way of talking does not therefore imply that physical things or events are eternal in the way in which the number 7 is'.[24] We must therefore reject Whitrow's odd claim that according to the relativistic conception of Minkowski, 'external events *permanently* exist and we merely come across them'.[25] According to Minkowski's conception, an event qualifies as a *becomingless* occurrence by occurring in a network of relations of earlier and later and thus can be said to occur 'at a certain time t'. Hence to assert tenselessly that an event exists (occurs) is to claim that there is a time or clock reading t with which it coincides. But surely this assertion does not entail the absurdity that the event exists (occurs) at *all* clock times or 'permanently'. To occur tenselessly at some time t or other is not at all the same as to exist 'permanently'.

Whitrow himself acknowledges Minkowski's earlier-later relations when he says correctly that 'the relativistic picture of the world recognizes only a difference between earlier and later and not between past, present, and future'.[26] But he goes on to query: 'if no events *happen*, except our observations, we might well ask—why are the latter exceptional?'[27] I reply first of all: But Minkowski asserts that events happen tenselessly in the sense of occurring at certain clock times. And as for the exceptional status of the events which we register in observational awareness, I make the following obvious but only partial retort: being registered in awareness, these events are *eo ipso* exceptional.

I say that this retort is only partial because behind Whitrow's question there lurks a more fundamental query. This query must be answered by those of us who claim with Russell that 'past, present, and future arise from time-relations of subject and object, while

---

24 J. J. C. Smart, *op. cit.*, p. 139.

25 G. J. Whitrow, *op. cit.*, p. 88, n. 2 (my italics).

26 *Ibid.*, p. 293.

27 *Ibid.*, p. 88, n. 2.

earlier and later arise from time-relations of object and object'.[28] That query is: Whence the becoming in the case of mental events that become and are causally dependent on physical events, given that physical events themselves do not become independently of being perceived but occur tenselessly? More specifically, the question is: if our *experiences* of (extra and/or intradermal) physical events are causally dependent upon these events, how is it that the former *mental* events can properly qualify as being 'now', whereas the eliciting physical events *themselves* do not so qualify, and yet both kinds of events are (severally and collectively) alike related by quasi-serial relations of earlier and later?[29]

But, as I see it, this question does not point to refuting evidence against the mind-dependence of becoming. Instead, its force is to demand (a) the recognition that the complex mental states of judgmental awareness as such have distinctive features of their own, and (b) that the articulation of these features as part of a theoretical account of 'the place of mind in nature' acknowledges *what may be peculiar to the time of awareness*. That the existence of features peculiar to the time of awareness does not pose perplexities militating against the mind-dependence of becoming seems to me to emerge from the following three counter questions, which I now address to the critics:

(1) Why is the mind-dependence of becoming more perplexing than the mind-dependence of common-sense color attributes? That is, why is the former more puzzling than that physical events like the reflection of certain kinds of photons from a surface causally induce mental events like seeing blue which are qualitatively fundamentally different in some respects? In asking this question, I am *not* assuming that nowness is a *sensory quality* like red or sweet, but only that nowness and sensory qualities alike depend on awareness.

(2) Likewise assuming the causal dependence of mental on physical events, why is the mind-dependence of becoming more puzzling than the fact that the raw feel components of mental events, such as a particular event of seeing green, are not members

---

[28] Bertrand Russell, 'On the Experience of Time', *The Monist*, 25 (1915), 212.

[29] The need to deal with this question has been pointed out independently by Donald C. Williams and Richard Gale.

of the *spatial* order of physical events?[30] Yet mental events and the raw feels ingredient in them are part of a time system of relations of earlier and later that comprises physical events as well.[31]

(3) Mental events must differ from physical ones in some respect qua being mental, as illustrated by their not being members of the same system of spatial order. Why then should it be puzzling that on the strength of the *distinctive* nature of conceptualized awareness and self-awareness, mental events differ further from physical ones with respect to becoming, while both kinds of events sustain temporal relations of simultaneity and precedence?

What is the reasoning underlying the critics' belief that their question has the capability of pointing to the refutation of the mind-dependence of becoming? Their reasoning seems to me reminiscent of Descartes' misinvocation of the principle that there must be nothing more in the effect than is in the cause *à propos* of one of his arguments for the existence of God: the more perfect, he argued, cannot proceed from the less perfect as its efficient and total cause. The more perfect, i.e., temporal relations involving becoming, critics argue, cannot proceed from the less perfect, i.e., becomingless physical time, as its efficient cause. By contrast, I reason that nowness (and thereby pastness and futurity) are features of events *as experienced* conceptually, *not* because becoming is likewise a feature of the physical events which causally elicit our awareness of them, but because these elicited states are indeed specified states of *awareness*. Once we recognize the role of awareness here, then the diversity and order of the events of which we have awareness in the form of now-contents gives rise to the transiency of the now as explained in Section III above, due cautions being exercised, as I emphasized there, that this transiency not be construed tautologically.

In asserting the mind-dependence of becoming, I allow fully that the kind of neurophysiological brain state which underlies our

---

[30] Mental events, as distinct from the neurophysiological counterpart states which they require for their occurrence, are *not* in our heads in the way in which, say, a biochemical event in the cortex or medulla oblongata is.

[31] Thus a conscious state of elation induced in me by the receipt of good news from a telephone call C_1 could be *temporally between* the physical chain C_1 and another such chain C_2 consisting of my telephonic transmission of the good news to someone else.

mere awareness of an event as simply occurring now differs in specifiable ways from the ones underlying tick-tock or melody awareness, memory-awareness, anticipation-awareness, and dream-free sleep. But I cannot see why the states of awareness which make for becoming must have physical event-counterparts which isomorphically become in their own right. Hence I believe to have coped with Whitrow's question as to why only perceived events become. Indeed, it seems to me that the thesis of mind-dependence is altogether free from an important perplexity which besets the opposing claim that physical events are inherently past, present, and future. This perplexity was stated by Smart as follows: 'If past, present, and future were real properties of events [i.e., properties possessed by physical events independently of being perceived], then it would require [non-trivial] explanation that an event which becomes present [i.e., qualifies as occurring *now*] in 1965 becomes present [now] at that date and not at some other (and this would have to be an explanation over and above the explanation of why an event of this sort occurred in 1965) '.[32] It would, of course, be a complete trivialization of the thesis of the mind-*in*dependence of becoming to reply that *by definition* an event occurring at a certain clock time *t* has the unanalyzable attribute of nowness at time *t*.

Thus to the question 'Whence the becoming in the case of mental events that become and are causally dependent on physical events which do not themselves become?' I reply: 'Becoming can characterize mental events qua their being both bits of *awareness* and sustaining relations of temporal order'.

The awareness which each of several human percipients has of a given physical event can be such that all of them are alike prompted to give the same tensed description of the external event. Thus, suppose that the effects of a given physical event are simultaneously registered in the awareness of several percipients such that they each perceive the event as occurring at essentially the time of their first awareness of it. Then they may each think at that time that the event belongs to the present. The parity of access to events issuing in this sort of intersubjectivity of tense has prompted the common-sense belief that the nowness of a physical event is an intrinsic, albeit transient attribute of the event. But this kind of

---

32 J. J. C. Smart, *op. cit.*, p. 135.

intersubjectivity does not discredit the mind-dependence of becoming; instead, it serves to show that the becoming present of an event, though mind-dependent no less than a pain, need not be *private* as a pain is. Some specific person's particular pain is private in the sense that this person has privileged access to its raw feel component.[33] The mind-dependence of becoming is no more refuted by such intersubjectivity as obtains in regard to tense than the mind-dependence of common-sense color attributes is in the least disproven by agreement among several percipients as to the color of some chair.

VI. BECOMING AND THE CONFLICT BETWEEN
DETERMINISM AND INDETERMINISM

If the doctrine of mind-dependence of becoming is correct, a very important consequence follows, which seems to have been previously overlooked: Let us recall that the nowness of events is generated by (our) conceptualized *awareness* of them. Therefore, *nowness is made possible by processes sufficiently macro-deterministic (causal) to assure the requisitely high correlation between the occurrence of an event and someone's being made suitably aware of it.* Indeed, the very concept of experiencing an external event rests on such macro-determinism, and so does the possibility of empirical knowledge. In short, insofar as there is a transient present, it is made possible by the existence of the requisite degree of macro-determinism in the physical world. And clearly, therefore, the transiency of the present can obtain in a completely deterministic physical universe, be it relativistic or Newtonian.

The theory of relativity has repudiated the uniqueness of the simultaneity slices within the class of physical events which the Newtonian theory had affirmed. Hence Einstein's theory certainly precludes the conception of 'the present' which some defenders of the objectivity of becoming have linked to the Newtonian theory. But it must be pointed out that the doctrine of the mind-dependence of becoming, being entirely compatible with the Newtonian

---

[33] I am indebted to Richard Gale for pointing out to me that since the term 'psychological' is usefully reserved for mind-dependent attributes which are private, as specified, it would be quite misleading to assert the mind-dependence of tense by saying that tense is 'psychological'. In order to allow for the required kind of intersubjectivity, I have therefore simply used the term 'mind-dependent'.

theory as well, does not depend for its validity on the espousal of Einstein's theory as against Newton's.

Our conclusion that there can be a transient now in a completely deterministic physical universe is altogether at variance with the contention of a number of distinguished thinkers that the indeterminacy of the laws of physics is both a necessary and sufficient condition for becoming. And therefore I now turn to the examination of their contention.

According to such noted writers as A. S. Eddington, Henri Bergson, Hans Reichenbach, H. Bondi, and G. J. Whitrow, it is a distinctive feature of an *indeterministic* universe, as contrasted with a deterministic one, that physical events belong to the present, occur *now*, or come into being over and above merely becoming present in *awareness*. I shall examine the argument given by Bondi, although he no longer defends it, as well as Reichenbach's argument. And I shall wish to show the following: insofar as events do become, the indeterminacy of physical laws is neither sufficient nor necessary for conferring nowness or presentness on the occurrences of events, an attribute whereby the events come into being. And thus my analysis of their arguments will uphold my previous conclusion that far from depending on the indeterminacy of the laws of physics, becoming requires a considerable degree of macro-determinism *and* can obtain in a completely deterministic world. Indeed, I shall go on to point out that not only the becoming of any kind of event but the temporal order of earlier and later among physical events depends on the at least quasi-deterministic character of the macrocosm. And it will then become apparent in what way the charge that a deterministic universe must be completely *timeless* rests on a serious misconstrual of determinism.

Reichenbach contends: 'When we speak about the progress of time [from earlier to later] . . . , we intend to make a synthetic [i.e., factual] assertion which refers both to an immediate experience and to physical reality'.[34] And he thinks that this assertion about events coming *into* being independently of mind—as distinct from merely occurring tenselessly at a certain clock time—can be justified in regard to physical reality on the basis of indeterministic

---

[34] Hans Reichenbach, *The Philosophy of Space and Time* (New York: Dover Publications, Inc., 1958) , pp. 138–39.

quantum mechanics by the following argument:[35] In classical deterministic physics, both the past and the future were determined in
relation to the present by one-to-one functions even though they
differed in that there could be direct observational records of the
past and only predictive inferences concerning the future. On the
other hand, while the results of past measurements on a quantum
mechanical system are *determined* in relation to the present *records*
of these measurements, a present measurement of one of two conjugate quantities does *not* uniquely determine in any way the result
of a *future* measurement of the other conjugate quantity. Hence,
Reichenbach concludes:

> The concept of "becoming" acquires significance in physics: the
> present, which separates the future from the past, is the moment
> at which that which was undetermined becomes determined, and
> "becoming" has the same meaning as "becoming determined." . . . it
> is with respect to "now" that the past is determined and that the
> future is not.[36]

I join Hugo Bergmann[37] in rejecting this argument for the following reasons. In the indeterministic quantum world, the relations
between the sets of measurable values of the state variables characterizing a physical system at different times are, in principle, *not* the
one-to-one relations linking the states of classically behaving closed
systems. But I can assert correctly in 1966 that this holds for a given
state of a physical system and its absolute future quite independently of whether that state occurs at midnight on December 31, 1800 or
at noon on March 1, 1984. Indeed, if we consider *any one* of the
temporally successive regions of space-time, we can veridically assert
the following at *any* time: the events belonging to that particular
region's absolute past could be (more or less) uniquely specified in
records which are a part of that region, whereas its particular
absolute future is thence quantum mechanically unpredictable. Accordingly, *every* event, be it that of Plato's birth or the birth of a
person born in the year 2000 A.D., *at all times* constitutes a divide

[35] Hans Reichenbach, 'Les Fondements Logiques de la Mécanique des Quanta',
Annales de l'Institut Poincaré, 13 (1953) , 154–57.

[36] *Ibid.*

[37] Cf. H. Bergmann, *Der Kampf um das Kausalgesetz in der jüngsten Physik*
(Braunschweig: Vieweg & Sohn, 1929) , pp. 27–28.

BASIC ISSUES IN THE PHILOSOPHY OF TIME

in Reichenbach's sense between its own recordable past and its unpredictable future, *thereby satisfying Reichenbach's definition of the 'present' or 'now' at any and all times!* And if Reichenbach were to reply that the indeterminacies of the events of the year of Plato's birth have already been transformed into a determinacy, whereas those of 2000 A.D. have not, then the rejoinder would be: this tensed conjunction holds for any state between sometime in 428 B.C. and 2000 A.D. that qualifies as now during that interval on grounds other than Reichenbach's asymmetry of determinedness; but the second conjunct of this conjunction does not hold for any state after 2000 A.D. which qualifies as now after that date. Accordingly, contrary to Reichenbach, the now of conceptualized awareness must be invoked tacitly at time t, if the instant t is to be nontrivially and nonarbitrarily singled out as present or now by Reichenbach's criterion, i.e., if the instant t is to be uniquely singled out at time t as being 'now' in virtue of being the threshold of the transition from indeterminacy to determinacy.

Turning to Bondi, we find him writing:

> . . . the flow of time has no significance in the logically fixed pattern demanded by deterministic theory, time being a mere coordinate. In a theory with indeterminacy, however, the passage of time transforms statistical expectations into real events.[38]

If Bondi intended this statement to assert that the indeterminacy makes for our human inability to know in advance of their actual occurrence what particular kinds of events will in fact materialize, then, of course, there can be no objection. For in an indeterministic world, the attributes of specified kinds of events are indeed not uniquely fixed by the properties of earlier events and are therefore correspondingly unpredictable. But I take him to affirm beyond this the following traditional philosophical doctrine; in an indeterministic world, events come *into* being by becoming present with time, whereas in a deterministic world the status of events is one of merely occurring tenselessly at certain times. And my objections to his appeal to the transformation of statistical expectations into real events by the passage of time fall into several groups as follows.

(1) Let us ask: what is the character and import of the differ-

---

38 H. Bondi, 'Relativity and Indeterminacy', *Nature,* 169 (1952) , 660.

ence between a (micro-physically) indeterministic and a deterministic physical world in regard to the attributes of future events? The difference concerns only the type of functional connection linking the attributes of future events to those of present or past events. Thus, *in relation to the states existing at other times,* an indeterministic universe allows alternatives as to the attributes of an event that occurs at some given time, whereas a deterministic universe provides no corresponding latitude. But this difference does *not* enable (micro-physical) indeterminism—as contrasted with determinism—to make for a difference in the *occurrence-status* of future events by enabling them to come *into* being. Hence in an indeterministic world, physical events no more *become* real (i.e., present) and are no more precipitated into existence, as it were, than in a deterministic one. In either a deterministic or indeterministic universe, events can be held to come into being or to become 'actual' by becoming *present in (our) awareness;* but becoming actual in virtue of occurring *now* in that way no more makes for a mind-independent coming into existence in an indeterministic world than it does in a deterministic one.

(2) Nor does indeterminacy as contrasted with determinacy make for any difference whatever at any time in regard to the *intrinsic attribute-specificity* of the future events themselves, i.e., to their being (tenselessly) what they are. For in either kind of universe, it is a fact of logic that what will be, will be, no less than what is present or past is indeed present or past![39] The result of a future quantum mechanical measurement may not be definite prior to its occurrence in relation to earlier states, and thus our prior knowledge of it correspondingly cannot be definite. But a quantum mechanical event has a tenseless occurrence status at a certain time which is fully compatible with its intrinsic attribute-definiteness just as a measurement made in a deterministic world does. Contrary to a widespread view, this statement holds also for those events which are constituted by energy states of quantum mechanical

---

[39] I am indebted to Professor Wilfrid Sellars for having made clarifying remarks to me in 1956 which relate to this point. And Costa de Beauregard has reminded me of the pertinent French dictum *Ce qui sera, sera.* There is also the well-known (Italian) song *Che Sera, Sera.*

systems, since energy *can* be measured in an arbitrarily short time in that theory.[40]

Let me remark parenthetically that the quantum theory of measurement has been claimed to show that the *consciousness* of the human observer is essential to the definiteness of a quantum mechanical event. I am not able to enter into the technical details of the argument for this conclusion, but I hope that I shall be pardoned for nonetheless raising the following question in regard to it. Can the quantum theory account for the relevant physical events which presumably occurred on the surface of the earth *before* man and his consciousness had evolved? If so, then these physical events cannot depend on human consciousness for their specificity. On the other hand, if the quantum theory cannot in principle deal with *pre*-evolutionary physical events, then one wonders whether this fact does not impugn its adequacy in a fundamental way.

In an indeterministic world, there is a lack of attribute-specificity of events *in relation to events at other times*. But this *relational* lack of attribute-specificity cannot alter the fact of logic that an event is intrinsically attribute-specific in the sense of tenselessly being what it is at a certain clock time t.[41]

It is therefore a far-reaching mistake to suppose that unless and until an event of an indeterministic world belongs to the present or past, the event must be *intrinsically* attribute-*in*definite. This error

---

40 Yakir Aharonov and David Bohm have noted that time does not appear in Schrödinger's equation as an operator but only as a parameter and have pointed out the following: (1) The time of an energy state is a dynamical variable belonging to the measuring apparatus and therefore *commutes* with the energy of the observed system. (2) Hence the energy state and the time at which it exists do *not* reciprocally limit each other's well-defined status in the manner of the noncommuting conjugate quantities of the Heisenberg Uncertainty Relations. (3) Analysis of illustrations of energy measurement (e.g., by collision) which seemed to indicate the contrary shows that the experimental arrangements involved in these examples did not exhaust the measuring possibilities countenanced by the theory. Cf. their two papers on 'Time in the Quantum Theory and the Uncertainty Relation for Time and Energy', *Physical Review*, 122 (1961), 1649, and *Physical Review*, 134 (1964), B1417. I am indebted to Professor A. Janis for this reference.

41 A helpful account of the difference relevant here between being *determinate* (i.e., intrinsically attribute-specific) and being *determined* (in the relational sense of causally necessitated or informationally ascertained), is given by Donald C. Williams in *Principles of Empirical Realism* (Springfield, Ill.: Charles C Thomas, 1966), pp 274 ff.

is illustrated by Capek's statement that in the case of an event 'it is
only its presentness [i.e., nowness] which creates its specificity . . .
by eliminating all other possible features incompatible with it'.[42]
Like Bondi, Capek overlooks that it is only with respect to some
now or other that an event can be future at all to begin with and
that the lack of attribute-specificity or 'ambiguity' of a future event
is not intrinsic but relative to the events of the prior now-perspec-
tives.[43] In an indeterministic world, an event is intrinsically attri-
bute-determinate by being (tenselessly) what it is (tenselessly),
regardless of whether the time of its occurrence be now (the present) or not. What makes for the coming into being of a future
event at a later time t is *not* that its attributes are indeterministic
with respect to prior times but only that it is registered in the now-
content of awareness at the subsequent time t.

(3) Two quite different things also seem to be confused when it
is inferred that in an indeterministic quantum world, future physical events themselves distinctively come into being with the passage
of time over and above merely occurring and becoming present to
awareness, whereas in a deterministic universe they do not come
into being: (i) the epistemic precipitation of the *de facto* event-
properties of future events out of the wider matrix of the possible
properties allowed in advance by the quantum-mechanical probabilities, a precipitation or becoming definite which is constituted by
our getting to *know* these *de facto* properties at the later times, and
(ii) a mind-independent coming into being over and above merely
occurring and becoming present to awareness at the later time. The
epistemic precipitation is indeed effected by the passage of time
through the transformation of a merely statistical expectation into a
definite piece of available information. But this does *not* show that
in an indeterministic world there obtains any kind of becoming
present ('real') with the passage of time that does not also obtain

---

[42] Capek, *op. cit.*, p. 340.

[43] Capek writes further: 'As long as the ambiguity of the future is a mere
appearance due to the limitation of our knowledge, the temporal character of the
world remains necessarily illusory', and 'the principle of indeterminacy . . . means
the *reinstatement of becoming in the physical world*' [*ibid.*, p. 334]. But granted
that the indeterminacy of quantum theory is ontological rather than merely
epistemological, this indeterminacy is nonetheless relational and hence unavailing
as a basis for Capek's conclusions.

in a deterministic one. And in either kind of world, becoming as distinct from mere occurrence at a clock time requires conceptualized awareness.

We see then that the physical events of the indeterministic quantum world as such do not come into being anymore than those of the classical deterministic world but alike occur tenselessly. And my earlier contention that the transient now is mind-dependent and irrelevant to physical events as such therefore stands.

Proponents of indeterminism as a physical basis of objective becoming have charged that a deterministic world is timeless. Thus, Capek writes:

> . . . the future in the deterministic framework . . . becomes something *actually* existing, a sort of disguised and hidden present which remains hidden only from our limited knowledge, just as distant regions of space are hidden from our sight. "Future" is merely a label given by us to the unknown part of the *present* reality, which exists in the same degree as scenery hidden from our eyes. As this hidden portion of the present is *contemporary* with the portion accessible to us, the temporal relation between the present and the future is eliminated; the future loses its status of "futurity" because instead of succeeding the present it *coexists* with it.[44]

In the same vein, G. J. Whitrow declares:

> There is indeed a profound connection between the reality of time and the existence of an incalculable element in the universe. Strict causality would mean that the consequences pre-exist in the premises. But, if the future history of the universe pre-exists logically in the present, why is it not already present? If, for the strict determinist, the future is merely "the hidden present," whence comes the illusion of temporal succession?[45]

But I submit that there is a clear and vast difference between the relation of one-to-one functional connection between two temporally-separated states, on the one hand, and the relation of temporal coexistence or simultaneity on the other. How, one must ask, does the fact that a future state is uniquely specified by a present state detract in the least from its being later and entail that it paradoxi-

---

[44] *Ibid.*, pp. 334–35, cf. also p. 164.

[45] G. J. Whitrow, *op. cit.*, p. 295.

cally exists at present? Is it not plain that Capek trades on an ambiguous use of the terms 'actually existing' and 'coexists' to confuse the time sequential relation of being *determined* by the present with the simultaneity relation of contemporaneity with the present? In this way, he fallaciously saddles determinism with entailing that future events exist now just because they are determined by the state which exists now. When he tells us that according to determinism's view of the future, 'we are already dead without realizing it now',[46] he makes fallacious use of the correct premiss that according to determinism, the present state uniquely specifies at what later time any one of us shall be dead. For he refers to the determinedness of our subsequent deaths misleadingly as our 'already' being dead and hence concludes that determinism entails the absurdity that we are dead *now*! Without this ambiguous construal of the term 'already', no absurdity is deducible.

When Whitrow asks us why, given determinism, the future is not already present even though it 'pre-exists logically in the present', the reply is: precisely because existing at the present time is radically different in the relevant temporal respect from what he calls 'logical pre-existence in the present'. Whitrow ignores the fact that states hardly need to be simultaneous just because they are related by one-to-one functions. And he is able to claim that determinism entails the illusoriness of temporal succession (i.e., of the earlier-later relations) only because he uses the term 'hidden present' just as ambiguously as Capek uses the term 'coexists'. But, more fundamentally, we have learned from the theory of relativity that events sustain time-like separations to one another *because* of their *causal* connectibility or deterministic relatedness, *not* despite that deterministic relatedness. And nothing in the relativistic account of the temporal order depends on the existence of an indeterministic microphysical substratum! Indeed, in the absence of the causality assumed in the theory in the form of causal (signal) connectibility, it is altogether unclear how the system of relations between events would possess the kind of *structure* that we call the 'time' of physics.[47]

---

46 M. Capek, *op. cit.*, p. 165.

47 Accordingly, we must qualify the following statement by J. J. C. Smart, *op. cit.*, pp. 141–42: 'We can now see also that the view of the world as a space-time

VII.   SUMMARY

In this lecture, I have presented my reasons for denying that now-ness and temporal becoming are entitled to a place within physical theory, be it deterministic or indeterministic. On the other hand, the temporal relations of earlier than, later than, and of simulta-neity do, of course, obtain among physical events in their own right in the sense familiar from the theory of relativity. Hence, if the 'meaning' of time is held to comprise becoming or passage, then one of the features of time is mind-dependent. But in characterizing becoming as mind-dependent, I allow fully that the mental events on which it depends themselves require a biochemical physical base or possibly a physical basis involving cybernetic hardware.

ADOLF GRÜNBAUM

UNIVERSITY OF PITTSBURGH

---

manifold no more implies determinism than it does the fatalistic view that the future "is already laid up". It is compatible both with determinism and with indeterminism, i.e., both with the view that earlier time slices of the universe are determinately related by laws of nature to later time slices and with the view that they are not so related'. This statement needs to be qualified importantly, since it would not hold if 'indeterminism' here meant a macro-indeterminism such that macroscopic causal chains would not exist.

For a discussion of other facets of the issues here treated by Smart, see A. Grünbaum, "Free Will and Laws of Human Behavior," *The American Philosophi-cal Quarterly*, October 1971.

INDEX